REGENERATION

REGENERATION

The Rescue *of a* Wild Land

ANDREW PAINTING

BIRLINN

First published in 2021 by
Birlinn Limited
West Newington House
10 Newington Road Edinburgh
EH9 1QS

www.birlinn.co.uk

Copyright © Andrew Painting 2021
The map on pp. x–xi contains Ordnance Survey Data
Crown Copyright and Database Right 2020

The moral right of Andrew Painting to be identified as the author of this work has been asserted by him in accordance with the Copyright, Designs and Patents Act 1988

All rights reserved. No part of this publication may be reproduced, stored or transmitted in any form without the express written permission of the publisher.

ISBN: 978 1 78027 714 1

British Library Cataloguing-in-Publication Data
A catalogue record for this book is available from the British Library

Typeset by Initial Typesetting Services, Edinburgh

Printed and bound by Clays Ltd Elcograf S.p.A.

Contents

Prologue: The Nature of the Beast	vii
Map	x
Introduction: Three Trees	1
Part One: In the Woods	**23**
1. Scots Pine	37
2. Green Shield-moss	53
3. Roe Deer	63
4. Woodland Grouse	73
5. Kentish Glory	83
6. On Patrol	95
Part Two: On the Moors	**105**
7. Red Deer	119
8. Sphagnum Moss	130
9. Atlantic Salmon	144
10. Curlew	157
11. Hen Harrier	171

Part Three: In the Mountains   **187**

12. Alpine Sow-thistle   197

13. Sphinx   207

14. Footpath   217

15. Dotterel   226

16. Downy Willow   236

Epilogue: You Could Have It So Much Better   249

Afterword: The Thin Green Line   267

Notes and References   271

Index   295

Prologue: The Nature of the Beast

> Landscape is a work of the mind. Its scenery is built up as much from strata of memory as from layers of rock.
>
> Simon Schama, *Landscape and Memory*, 1996

We do not come to nature as impartial observers. We all have an emotional and cultural attachment to all the life with which we share the earth, from the smallest weed to the tallest tree. Such experiences and interests have the power to enrich and enhance our relationships with nature. They also have the power to complicate and destroy them.

This is worth remembering as we explore stories of the conservation efforts underway at Mar Lodge Estate National Nature Reserve. With its ancient pinewoods, subarctic mountains and rolling bogs and moors, the estate sits at a complex and often awkward juncture between these real and imagined landscapes. It is a deeply beautiful place, with the power to affect humans in ways that few other British landscapes can. It is a place of soaring eagles and roaring stags, a refuge for some of the rarest creatures in Scotland, a place of deadly cold, avalanches and remoteness, where the redstart still sings in the spring birch budburst and the salmon still runs up cascades. It is an ancient landscape which provides us with links to our past and refuge from our present. But it is also a contested place, which has been damaged by conflicts that blight the nature of Scotland and continue to divide its people.

In recent years it has become a place of cooperation and compromise. It is cared for by a bewildering array of people. This book is about that work, the people who do it, and the creatures that inspire them to do it. Their work is not easy. In fact, it is often extremely difficult. It has led to acrimony and hostility. It remains controversial, difficult, hard to pin down and understand. Yet this book is a collection of stories of regeneration, redemption and reconnection, of chances taken to fight the destruction of the environment, of collaboration and communication between different groups and cultures to make a better future for everyone. These are stories of people from all walks of life coming together and striving forwards to a new, biodiverse future, where humans can forge new relationships with nature, in new, enriched landscapes of the mind. In a world of mass extinction, these are rare stories of hope.

The whole, unabridged story of Mar Lodge Estate stretches back 10,000 years and more. It is a story involving queens and kings, princesses, dukes, earls, hunters, Jacobites, crofters, scientists, poets, adventurers. This is the place where a young Byron nearly met his end, and where Nan Shepherd was inspired to write her greatest works. This land has been buffeted by centuries of human use and abuse, by grand tides of history and Events with a capital E, by wars and Clearances, by lawsuits and climate change. Events have a great bearing on this story. But it is not always grand actions and heroic deeds which make the world a better place. It is often the gradual accumulation of small acts of kindness and curiosity and skill. So this book is filled with these little events: a person looking for wood ants, another counting trees, another stalking a deer. These little events, over time, become big events, gain traction, then movement, and slowly the world becomes a better place.

The catalyst that set many (but by no means all) of these remarkable little events in motion happened in 1995, when the National Trust for Scotland, with significant help from a number of people and organisations, acquired 30,000 hectares of some of our most celebrated land. It was to be held in trust for the benefit of the nation.

This book is a celebration of that Event. It tells some stories from the front line of environmental conservation. I've been on the scene in a minor capacity at Mar Lodge for the last few years. I've been given a ringside seat to some of the most exciting, progressive conservation stories currently playing out in Britain. The future is far from certain, and our environment faces greater challenges than ever before, but these stories show that there is still hope for the future.

Mostly though, this book is my way of answering a thorny question which has dogged my career. It is posed at parties, social events, any time I have met someone new, or even reunited with someone I've not seen for a few years. It is an annoying question because to answer it properly takes slightly longer than the socially acceptable amount of time that one has to answer a question about one's job.

'What does an ecologist actually *do*?'

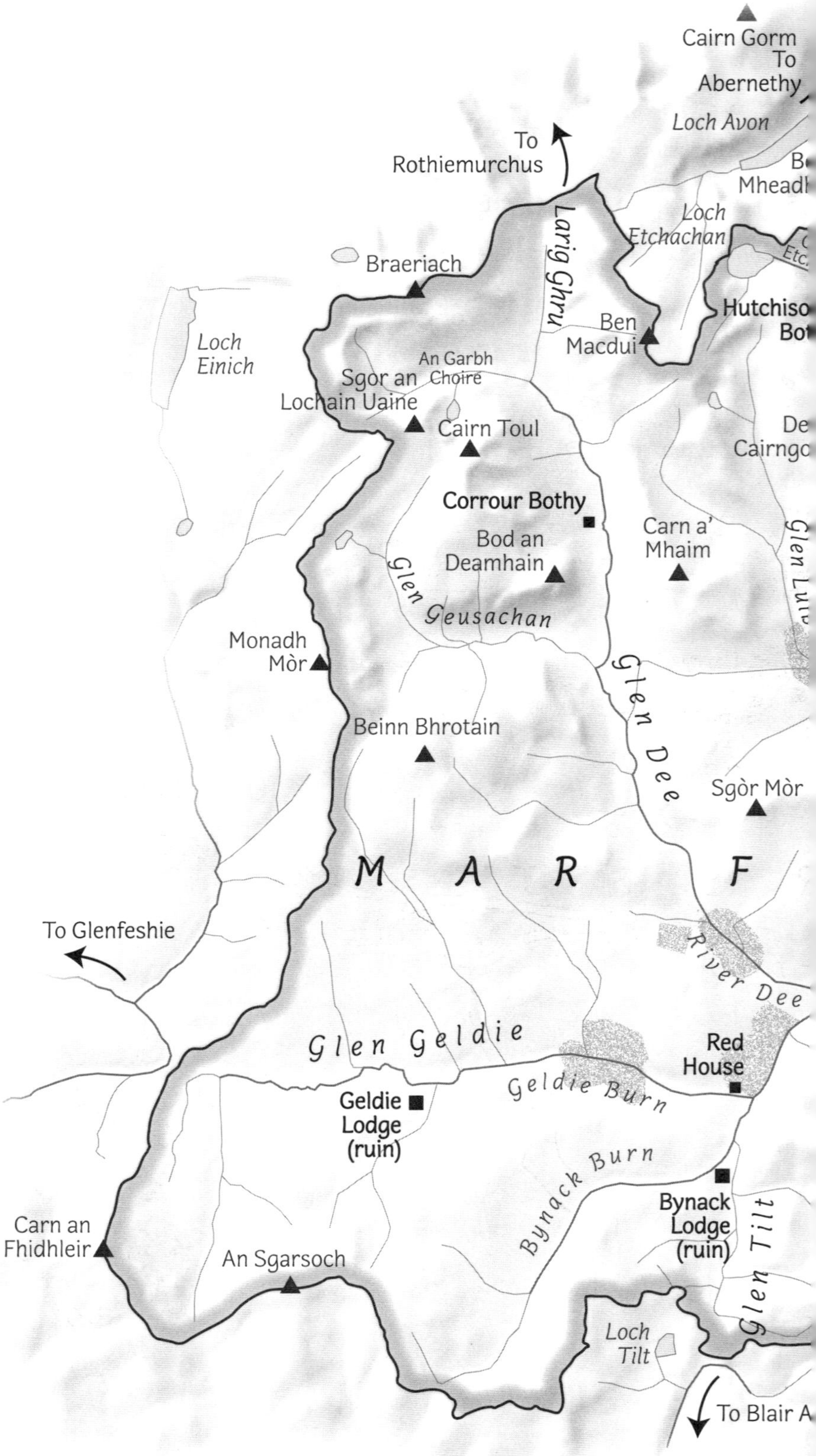
Cairn Gorm
To
Abernethy
Loch Avon
To
Rothiemurchus
Larig Ghru
Loch
Etchachan
Braeriach
Ben
Macdui
Hutchiso
Bo
Loch
Einich
An Garbh
Choire
Sgor an
Lochain Uaine
Cairn Toul
Corrour Bothy
Carn a'
Mhaim
Bod an
Deamhain
Glen
Geusachan
Monadh
Mòr
Glen Dee
Beinn Bhrotain
Sgòr Mòr
M A R
To Glenfeshie
River Dee
Glen Geldie
Red
House
Geldie Burn
Geldie
Lodge
(ruin)
Bynack Burn
Bynack
Lodge
(ruin)
Glen Tilt
Carn an
Fhidhleir
An Sgarsoch
Loch
Tilt
To Blair A

0 5 km
0 3 miles
N

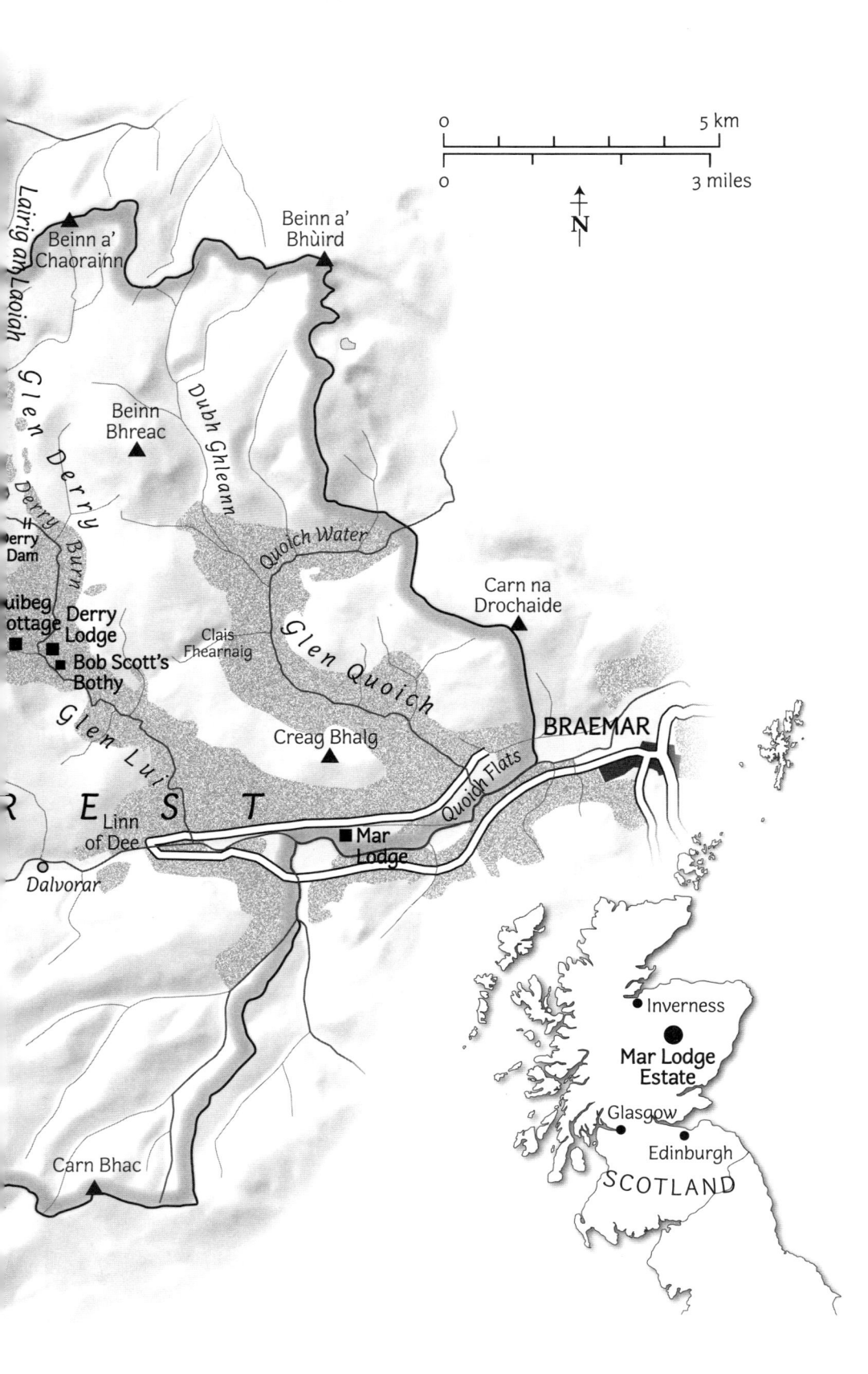

Lairig an Laoigh
Beinn a' Chaorainn
Beinn a' Bhùird
Glen Derry
Beinn Bhreac
Dubh Ghleann
Derry Burn
Derry Dam
Luibeg Cottage
Derry Lodge
Bob Scott's Bothy
Quoich Water
Clais Fhearnaig
Carn na Drochaide
Glen Quoich
Glen Lui
Creag Bhalg
BRAEMAR
Quoich Flats
FOREST
Linn of Dee
Mar Lodge
Dalvorar
Carn Bhac

Inverness
Mar Lodge Estate
Glasgow
Edinburgh
SCOTLAND

Introduction: Three Trees

> For words, like Nature, half reveal
> And half conceal the Soul within.
>
> Alfred, Lord Tennyson, 'In Memoriam', 1849

May is the finest month in Scotland, and it is sacrilege to spend even one May day indoors. There are hundreds of thousands of trees on Mar Lodge Estate, and millions of seedlings and saplings. But there are three particularly interesting specimens, and I'm setting out on a minor pilgrimage to see them.

It's a cold, clear, still day. I've started early because it is a very long walk. I leave from the old stables, eaves thronged with house martin nests, walk out past the blackcap singing in the willow by the burn, and head along the back road. I walk past Mar Lodge itself. It is a grand, imposing, red granite, red-tiled mansion, with two large wings stretching out east and west. It is not as old as it looks – Queen Victoria laid the first stone in 1895. The wildlife is in overdrive at this time of day. A brown hare lopes across the lawn, a red squirrel scampers up an old pine, its claws making a distinctive pitter-patter against the pine bark. I take a turning off into Doire Bhraghad,[*][1] the

---

[*] Most of the place names at Mar Lodge have Gaelic roots. I've provided rough translations to English where useful for the reader. For a fuller treatment of the place names of Mar Lodge, I'll push you in the direction of Adam Watson and Elizabeth Allen's *The Place Names of Upper Deeside* (Rothersthorpe: Paragon, 1984).

wooded brae, the Mar Forest. It's open woodland, grand old pines and birches, interspersed with thousands of young trees crowding the path. A mile or so on I surprise a black grouse. Unlike red grouse they don't call out your rudeness with an affronted 'go-back go-back'. But they do still make a meal of their departure, all ruffled feathers and hurt pride. It gives me pause to stop and listen. It's a gorgeous morning, and the clear air is full of song; willow warblers, tree pipits, redstarts, tits, chaffinches, crossbills, siskins, cuckoos. There's a great spotted woodpecker calling, and a minute after I hear a green woodpecker, which is something of a novelty this far north.

May is the finest month in Scotland because it combines the best weather with the most wildlife. Sometimes you can walk around the Cairngorms without seeming to find any wildlife at all. But not on a fine day in May. It is a strange thing in Britain to come across a landscape that is literally full of wildlife, and yet here it is.

Mar Lodge Estate sits in the heart of the Cairngorms, the mountains in the middle of Scotland. It is a two-hour drive from Edinburgh in the south and from Inverness in the north. The area covers the land to the west of the village of Braemar, renowned for its Highland games, and to the east of Glen Tilt, from the waters of the Dee to the peak of Ben Macdui.

It is a place of extremes and contradictions. It forms part of the largest stretch of high subarctic ground in Britain. It is the largest National Nature Reserve in Scotland; 50 per cent larger than the next biggest and 3,500 times larger than the smallest. It sits slap bang in the middle of the Cairngorms National Park, the largest national park in Britain. It takes three days to walk the estate's march, the boundary, if you're fit. The land boasts fifteen Munros (hills over 3,000 feet/914 metres) and four of the five highest mountains in Scotland. It has the highest source of any river in Britain. It is the coldest, windiest, snowiest place in Scotland. It holds its snow all year round, and snow can fall in any month of the year. It has eleven national and international environmental designations, a bewildering array of initialisms and acronyms, SPA, SAC, NNR,

SSSI, SAM, NSA, WLA, GCR, designed to protect it from the worst in ourselves. Over 5,000 species have been recorded on the estate, well over 10 per cent of all the species found in Scotland. It is an ancient landscape of granite hewn over hundreds of millions of years, hunched over by time and hollowed out by ice, but it is also a landscape of seasonal renewal and growth. It is wild land, often inaccessible, yet it has as rich a human history as anywhere in Scotland. It is thrumming with wildlife, and yet its soils are thin and acidic and spent. It is home to some of Scotland's most charismatic creatures but is missing many more species which were driven to extinction in Scotland by humans. It has suffered at our hands for centuries, yet remains one of the best places in Scotland for wildlife.

And here, in the Mar Forest, on a beautiful early morning in May, it is full. And here is our first specimen – the second largest recorded Scots pine in Scotland.

Scots pines are the trees of the Cairngorms. And what trees! There are few greater pleasures in life than walking through a pinewood after a downpour, breathing in the resinous smell and listening to the chaffinch's tinkling, jangling song, knowing that you can walk for an hour and not reach the limit of the wood. To see a glen draped in thousands of giant, ancient pines, with their shaggy, gingery red bark and dark green needles, is to see one of the last great natural spectacles afforded to us in the UK.

In 1993, the renowned Cairngorms scientist Dr Adam Watson sat at the foot of this particularly magnificent granny pine and was interviewed for the TV show *Scottish Eye*. Watson had spent his life studying, living with, and striding around the Cairngorms. As a child he had explored Mar Lodge's woods and mountains. He had worked here as a ghillie for a spell in the 1950s. He knew the ground as well as anyone alive. By the 1990s he was known as Mr Cairngorms, a title that suited his somewhat wild appearance, with his long ginger-and-white beard and loping gait. He was also known as a thorn in the side of the Establishment.

The heather and blaeberry around him were clipped low. A dead deer calf that had starved that winter was decomposing by the tree. What concerned Watson was that there were no young trees. They had all been eaten by deer. In fact, almost all of the young trees that had been produced by the Mar Lodge pines for the last two hundred years had been eaten by deer. Watson told the TV crew that the greatest challenge for wildlife in the Highlands was an overabundance of red deer, which was suppressing woodlands, causing them to age, and die, and not be replaced by younger trees. 'It's an old place,' he said, in his soft Aberdonian accent, 'but it's a decaying place.' He argued that big Highland sporting estates which manage the land so that people could shoot deer for sport were destroying the land for the rest of us. He pointed out that after centuries of forestry operations, the number of deer at Mar Lodge was far higher than what the already depleted woodland could support. In fact, there were more deer on the land than at any point since the last Ice Age.[2] If nothing was done, he said, the Caledonian pinewoods, which always had been a part of the Scottish landscape and culture, would be lost.[†]

In 2011, volunteers with the Woodland Trust made the same journey to the big tree. They were the ones who measured its girth and found it to be the second largest recorded Scots pine in Scotland at 6.09 metres.[3] Look at the photos they took, and you'll see that the landscape is similar to how Watson had found it eighteen years earlier, all low clipped heather and blaeberry, and no young trees.

Change happens very slowly and very quickly in woodlands. The tree is a fair yomp from the path, through thickets of birch and pine and waist-high heather. I startle pearl-bordered fritillaries into flight, but it's still cold and they're not quite ready to go yet. The tree

---

[†] Adam Watson looms large over this book. He was a prolific writer and researcher. A quick look through the references of this book will show you how influential his decades of research have been on our understanding of the nature of the Cairngorms.

has grown in girth no more than 5 centimetres further since 2011. Two enormous limbs have been lost, each the size of a tree itself. But now the pine is surrounded by new growth: young birches, rowans, willows and pines crowd for attention, from knee height to twice my height.

The tree is a monster. It's not as old as people think – more than two centuries old, but not much more than that. Its remarkable size is actually a product of human history: for the last 200 years, high numbers of red deer, managed for sport, have browsed away other trees that might have grown up to compete with it. Its size will ultimately prove to be its undoing, as it has made it unstable in the regular high winds that blow up along Deeside from Atholl. Sitting on a broken limb worn smooth by previous sitters, I look around at the explosion of life. Spring comes late to the Cairngorms, and the birches are not quite in full leaf yet. But one day soon they will be, all of them, all of a sudden, just like that. Under a bough, I see a rib from a red deer calf, and the find is just a tad on the nose for my taste.

I could write that what has emerged at Mar Lodge is a demonstration project of what can be achieved when groups with different interests and different ideas about our landscapes work together, somewhat harmoniously, for the common good. But, as we shall see, it is fairer to say that the truth of the matter is much more complicated than that.

It's still cold, and I have many miles to walk, so I move on, deeper through the Mar Forest.

The woodland opens up a couple of kilometres away at Black Bridge, which is actually a red bridge. Glen Lui, whose flanks were cleared of Caledonian pinewood in the eighteenth century, is the gateway to the southern Cairngorms. It now appears as a wilderness, but like all the Cairngorms glens its human history is as old as the pinewoods themselves. The fertile floor, now a lush green wildflower meadow, was under the crofter's plough for centuries. The footprints of their shielings and blackhouses litter the glen. The crofters were first cleared in 1726 to facilitate forestry operations, and again in 1840, to make way for deer to hunt. Humans and pinewoods have both flourished and suffered in equal measure in this glen.

Carn a' Mhaim and Derry Cairngorm, gatekeepers themselves to Ben Macdui, hove into view. There are snow patches on their eastern flanks, and I look forward to stomping through them in several hours' time. One time here I saw a juvenile golden eagle flying down the glen, attacked simultaneously by two buzzards, a kestrel and two hen harriers, but today I make do with just a buzzard. These largeish raptors, with their beautiful mewing call, were rare enough creatures themselves just a couple of decades ago. If they weren't competing with eagles for our cultural affections, we would think of them, perhaps, with the respect they deserve. As it is, I have started calling them 'justabuzzards', so regularly do I see them and think that they might be something rarer. Justabuzzards are a conservation success story: when we think of once-rare creatures as blithely as I think of justabuzzards, that is a sure sign of success.

I share the glen with two walkers, weighed down with large rucksacks, looking to spend a couple of days up here in the heart of things. They don't see the justabuzzard, or maybe they have and, like me, have dismissed it out of hand for its newfound ubiquity. There are tree plantations in the glen from in the 1970s. The Trust is working to make them less like plantations and more like native woodlands, removing non-native tree species, thinning clearings and increasing the amount of deadwood in them. It's a slow process, and for now they remain slightly incongruous square blocks of trees. Willows and birches are growing, sometimes densely, sometimes sparsely, along the length of the open glen, for the first time in perhaps two centuries. Above all else, though, my eyes are drawn to the young pines. They pack themselves densely along the open sandy soils of the burnsides. They cluster around the older trees and woodland edges. They grow strongly in some patches, weaker in wetter areas, between knee height and well over head height. They send scouts across the landscape, their silhouettes breaking the skyline. One way or another, they are everywhere in this glen.

I head past Derry Lodge, the grand hunting lodge, and into Glen Derry, which is as remarkable a glen as any in Scotland. Caledonian pinewood drapes the lower reaches of the glen for a couple of miles, obscuring the hills. A merry band of Scottish crossbills calls from the tops of the trees. At Derry Dam, the pines begin to thin out, a relic of the forestry operations of the eighteenth century. Here the glen is laid bare; a vast U-shaped valley, hewn out by a monstrous glacier over 10,000 years ago, bounded by four Munros. Four great corries sprawl along its western flank, while scree slopes slouch the length of the eastern wall – a giant drystane dyke. To the north sit the tors of Beinn Mheadhoin, the middle hill, and further still, the great forest of Abernethy.

The scree slopes are the reason I'm here. I climb up to an outlying cluster of pines, isolated from the edge of the wood. It's tough going in the rough scree, but I notice small, wind-stunted pine seedlings popping through the rocks at my feet. These are the latest trees in a very special lineage. These are the descendants of Mar Lodge's oldest known tree. There it is, a lone pine, gnarled and twisted, with a sparse canopy, set back even from the other members of this isolated group.

Were it not for the magnificent location, you'd be inclined to find this tree something of an anticlimax. But then, that is somewhat the point. Its location, cold and exposed and lacking in nutrients and firm anchorage, has stunted its growth, and in so doing allowed it to last to an older age than is usual for Scots pine. It is dated to at least 1477, but it probably started growing sometime in the 1450s, making it one of the two oldest known Scots pines in Scotland.[4] When Queen Victoria wrote about the 'fine firs' of Glen Derry in her diaries, it was already an exceptionally old tree.[5] It was old even when Glen Lui was cleared of people to make way for deer to hunt. It was getting on for three centuries old when the last wolves of Scotland were killed, and when tens of thousands of trees were lost from the glen in the forestry operations of the eighteenth century. It was two centuries old, and more, when 'Bobbing' John Erskine, 23rd Earl of Mar, raised the banner for the '15 rebellion, only to forfeit the estate and flee to France. It was a century old when Mary, Queen of Scots hunted in Mar and Atholl for deer and wolves.

Sitting under its ancient boughs, I expect to think lofty, Ozymandian thoughts about the transience of human life and the enduring reality of nature. Facing such exceptional age, with a tree which had seen so much of the comings and goings of humans for half a millennium and more, I expect to feel some communion with nature. As it is, I mostly feel cold and uncomfortable. I slide about on the scree, struggling to find a snug spot, and eventually make do with an awkward seat, one leg keeping me balanced, the other flopping around uselessly. So much for lofty thoughts.

In Scotland, the largest parcels of lands are sporting estates. In a British context, they are enormous. They are the principal reason that, in 2012, 50 per cent of Scotland's land had just 432 owners.[6] A single owner has a huge amount of influence, for better or worse, over how that land is managed. Mar Lodge Estate is just such an estate, and a big one at that. It is an ancient, royal forest – its land set aside for the joy of the hunt for a privileged elite.[‡] It was a place *foris*, subject to its own laws. As we shall see, this has had an enormous bearing on its history and how we understand the land to this day. For a thousand years and more, the estate has had a single owner at a time. This owner has changed with the political whims and caprices of the age, with the competence of the owners, and with their ability to produce heirs. But it has always had pedigree. The original earls of Mar are among the oldest lineage of nobility in the world. The fortunes of the people of the estate were largely dependent on whoever happened to own the land at the time.[7] The highest number of people living and working on Mar Lodge came at some point in the early

---

‡ A 'forest' refers to this mode of law, rather than to the presence of trees. This is why so many of Scotland's deer forests are, paradoxically, treeless.

eighteenth century. They were foresters and crofters. They would use the woods for fuel and construction, their cattle would roam through the pinewoods and summer high in the hills. Following the Jacobite rebellion of 1715, the Earl of Mar forfeited his land, and the estate eventually made its way into the hands of William Duff, later to become the Earl of Fife. The crofters were cleared, first, to make way for forestry operations, then to make way for deer to hunt. By the 1850s, the land was a playground for the aristocracy. Foresters and crofters were replaced with gamekeepers, paid to produce high numbers of game by whatever means necessary. Every year Queen Victoria would arrive at nearby Balmoral for the season. She would ride out and enjoy the Mar woods, already by this point depleted by centuries of felling for timber, and lament the loss of the pines of Creag Bad an Eàs, felled earlier that century and not regrown. Prince Albert would hunt stags on the Mar Lodge ground. In the late nineteenth century, following the marriage of the Earl of Fife to Princess Louise, first daughter of King Edward VII, it became a royal hunting lodge. In 1959, following the death of Princess Alexandra, the estate passed to a nephew, Captain Ramsay.§ In 1962, the majority of the estate was bought by the Panchaud brothers, Swiss hoteliers who thought Mar Lodge could be turned into a ski resort. It could not, they eventually realised, but not before they had bulldozed vehicle tracks into the heart of the mountains in preparation for ptarmigan shooters and downhill skiers who would never come. In the 1980s the land made its way into the hands of John Kluge, an American oil tycoon. By 1995, the land had been managed first and foremost as a Highland sporting estate for almost two centuries, while field sports had been undertaken on the land for getting on for a millennium. In that time the culture of Highland sport and gamekeeping had

---

§ To confuse matters, the neighbouring Mar Estate remains in royal hands. Mar Estate takes in the land between Braemar and Glen Ey and is managed as a sporting estate. It's a good spot for black grouse and curlew, among other things.

emerged across the Highlands, and people had become used to the ways of life of a Highland sporting estate.

Whoever the owner, and whatever their outlook and management of the land, gradually, over the centuries, the great pinewoods of the Quoich, the Derry, Luibeg and Glen Dee were reduced to remnants. The woods and scrub of the mountains and burns were lost, the bogs became eroded and scarred. So too went the people, who had eked fields out of the unforgiving soils, leaving behind them the ruins of townships and shielings, and no less than seven illicit whisky stills. So too went the big mammals, the lynx, the boar, the beaver, the wolf, lost either through the destruction of their habitat or at the hands of hunters. Myriad other species, too small to notice or record for posterity, were lost, and we'll never know for sure what they were. By the nineteenth century the last remaining large mammals were red deer, the monarchs of the glen. And they remained in their thousands.

The image of a stag against a heather-clad purple moor is iconic, totemic even, and people will travel from across the world to see it and to stalk stags and shoot grouse on the open hill. For many, Highland sporting culture, the deer, the grouse moors, is what makes the Highlands special and beautiful. But for many others, the grazed, burned, drained moors and dying woodlands are a landscape of environmental and cultural destruction. This is the clash that lies at the heart of environmental conservation in the Highlands.

The early 1990s were an interesting time for the Scottish landscape. A new environmental watchdog, Scottish Natural Heritage (SNH), had recently been formed and was keen to show its teeth.⁋ Conservationists, buoyed by successful woodland regeneration projects in places like Beinn Eighe, Creag Meagaigh and Abernethy, were keen to pile the pressure on what they saw as an intransigent sporting industry. Sporting landowners, feeling the heat, and

⁋ To make matters more confusing, in 2020 SNH undertook a major rebranding and so is now called NatureScot, which is apparently better.

considering themselves to be the true guardians of the countryside, argued that if the tree-huggers had their way there would be no deer left in Scotland, and that with them would go the centuries-old Highland sporting tradition, and with that, the jobs and money. It was in this febrile atmosphere that one of the most prestigious sporting estates in Scotland was put on the market.

In retrospect, it seems inevitable that a charity like the Trust would be a good fit for a controversial landscape at a controversial time. The Trust was already responsible for the protection of a fair chunk of the 'jewels in the crown' of the Highlands: Glencoe, Torridon, Ben Lawers, Kintail, St Kilda. The Trust saw the value of conservation for the sake of biodiversity and for the great benefits of functioning ecosystems and beautiful landscapes being open for everyone to enjoy. But the Trust could also be respectful of culture, history and cultural landscapes. To put it simply: less tree-hugger, more tartan and tweed.**

At the time, however, the purchase was far from certain. The problem was that even if the Trust wanted to purchase the estate, it couldn't afford it. This simple fact brought together a disparate group of people and organisations to put it into Trust hands. Mar Lodge's owner, John Kluge, was happy to sell the estate to the Trust at well below market value, a bargain price of £5.572 million. Scottish Natural Heritage were excited by the prospect of saving the Mar Lodge woodlands. They would enter into a 25-year management agreement with the Trust worth around £125,000 a year. The

---

** Others have put the often cosy arrangement between the Trust and the landed gentry in stronger terms. Andy Wightman, another thorn in the side of the Establishment, wrote in *The Poor Had No Lawyers* (Edinburgh: Birlinn, 2013) that 'Scottish landowners embraced the heritage industry as a means of justifying the existence of large estates, mansion houses and titles and monopolised control of the National Trust for Scotland until relatively recently'. Given that the Trust's president from 2003 to 2012 was the Duke of Buccleuch, at the time the largest landowner in Scotland, Wightman might well have a point.

trustees of the National Heritage Memorial Fund would provide £1.5 million for the purchase of the estate. The Heritage Lottery Fund provided £8.015 million to act as an endowment for future management, and a further £732,000 towards capital works to the Lodge and other areas in the surrounding policies.[8]

The final piece in the puzzle was a £4 million gift from the Easter Charitable Trust, an organisation that no one had heard of before, principally because it had been set up directly to help the Trust buy Mar Lodge Estate. This money came with three conditions, presented here in their entirety:

1. The Trust shall manage the Estate so as to conserve its valuable ecological and landscape features in harmony with its maintenance as a Highland Sporting Estate for so long as field sports remain legal.
2. It is intended to demonstrate that the practice of field sports can be reconciled with the Trust's statutory obligation to promote public access.
3. The part of the Estate lying to the south and west of the River Dee and comprising heather moorland shall be sensitively managed to promote its proper conservation in terms of grouse habitat, nature conservation and landscape.[9]

From these conditions, and the principles of management drawn up by the Trust at the same time, came the three pillars which have informed all management decisions since then. The land would be managed for environmental and cultural heritage conservation, Highland sport and open access for all. From the outset, whatever management the Trust undertook would have to keep happy SNH, the Easter Trust and the Trust's own 300,000 members. Tough gig.

There was much speculation at the time as to who the mysterious benefactor behind the Easter Trust might be. In 2002, it was made public that the secret donor was Ann Marie Salvesen, of the

Salvesen shipping family.[10] Ann Marie shunned the limelight, and devoted her time and money to charitable works. She was a keen birdwatcher who also left money to the RSPB in her will. She had the means to improve the lives of those around her and did so. On her death her brother, Andrew Salvesen OBE, took over responsibility for the Easter Trust. Andrew is a trustee of the Game and Wildlife Conservation Trust and retains an active interest in the management of Mar Lodge Estate.

Ann Marie's generosity facilitated the protection of some of Scotland's finest land for the benefit of the nation, and, as we shall see, the beginnings of the return of the land to something like its former glory. She ensured that there would be a continuity of sporting tradition and culture at Mar Lodge. She also forced together two groups with very different opinions about how Scotland's landscape should look.

One thing was immediately clear – the Trust was going to make some serious changes to the management of the land. An ambitious 200-year vision was drawn up. With 30,000 hectares to play with, the Trust would endeavour to recreate one huge, functioning ecosystem, where different habitats could blend into one another. In two centuries, Mar Lodge Estate would be a place where:

> Landscape and habitats have been restored, and woodland regenerated to a natural treeline below the montane plateaux, with grazing animals managed at sustainable levels enabling native species and semi-natural habitats to flourish ... Management of the land for field sports as well as the functions of a Highland sporting estate have been fully integrated with conservation aims and access.[11]

A 550-year-old tree, and a 200-year management vision. I could sit under this tree for a year and not get my head around such timescales. I am surrounded by young trees, themselves stunted, not much more than waist height, poking out of the scree. These trees are the direct descendants of the first pines to colonise the area some 8,000 years ago. These trees will have wandered, some near, some far, from their ancestors. Faced with such exceptional age, the two-century vision to which the Trust is working, which is now one-eighth completed, seems a rather quaint idea.

Then, realising how this whole glen was a pinewood not so long ago, and how close it has all come to destruction, and how arbitrary the reasons for both its destruction and its survival, and how thin the line of people striving to stop it from being destroyed, my stomach drops.

It's hard to say how long the Old Tree has left. It is already well past its sell-by date. I wonder whether I'll outlive it. A peregrine glides nonchalantly up the glen, scaring grouse as it goes, as peregrines have done here for thousands of years. One of my legs has gone numb. It's time to leave off and take in the airs of the high hills.

The Hutchison's Memorial Hut is remarkable for several reasons. One is its location, high up in Coire Etchachan, an amphitheatre of rock and snow. Another is that, like all Scottish bothies managed by the extraordinary Mountain Bothies Association, it is open to all and free to use, a haven high up in the Cairngorms. It is a good place to muster forces for the steep ascent to Loch Etchachan, and then on to our third tree of the day.

The hills have not always been open for all. The fight for access to the countryside was long and bitter. Back in 1995, much was made of the Trust's lauded open access policy – like all of the Trust's land, Mar Lodge was to be open for all to go where they pleased and when they pleased. Such forward thinking has been superseded by some of the most remarkable legislation of recent years, the Scottish Outdoor Access Legislation of 2003. This broadly allows anyone to go anywhere they like, so long as they behave responsibly. But the Trust's commitment to open access, the upkeep of footpaths and the protection of the wild-land qualities of the place remain undimmed.

The juniper is returning to this, the juniper corrie,[††] alongside cold and wind-stunted pines, rare mountain-specialist willows, northern blaeberry and tall herb communities. Keep an eye on this place, come back in 2050; I assure you it will be all the more remarkable. A wheatear alights a rock near the hut, singing its sweet but forgettable song. There's no time to linger here, for the peaks are calling, but it is worth reciting the lines by Coleridge that adorn the bothy wall:

So will I build my altar in the fields,
And the blue sky my fretted dome shall be,
And the sweet fragrance that the wild flower yields
Shall be the incense I will yield to Thee[12]

Up to Loch Etchachan, the highest large loch in Britain, sitting at Munro height, and largely frozen over. Up higher, around its own encircling cliffs, I crunch through snow patches, not because I have to, but because I find walking through snow in May to be among the greatest of life's pleasures. The next tree takes a bit of finding, but I have a grid reference. It is a rowan, the mountain ash, and it is, we think, the second highest-altitude tree in the country. I'm 1,096 metres above sea level, and I'm on the Cairngorm plateau. I've found it, which is good, because I was half expecting it to be buried in snow.

It's not exactly a tree. It is knee-high, heavily grazed by mountain hares, red deer and reindeer from the Cairngorms' semi-wild herd. It's at least twenty years old, and probably much older. But it's here, on the roof of Scotland, a subarctic landscape of rock, snow, ice, moss and lichen.

This is truly wild country now. Few species can survive up here, a handful of arctic birds and arctic-alpine plants and not much else. This is a place of roaring winds, white-out and ice. But today, under

---

[††] A possible translation of the Gaelic Coire Etchachan, favoured by Adam Watson, but no one's quite sure where the name comes from.

a blue sky, it is a tonic. And even here, a rowan has managed to make a living.

Rowans are ephemeral trees, often short-lived, but catholic in taste. They are extremely hardy but also delicious to herbivores, which means that in Scotland you tend to find them on remote cliff edges, seemingly growing out of rock like this one, far out of the reach of nibbling deer. But up here, way into the subarctic tundra, is really a bit much, even for a rowan. This is a special scrap of twigs, if only for its willingness to survive against all the odds. It is an erratic – a blip on the graph, noise in the dataset. It will never be joined by other trees, nor will it ever grow any higher than half a metre or so above the ground. Hundreds of metres below us is Loch Avon. To the west is Ben Macdui. I can see the sea from here, the Moray Firth and the North Sea. I'm reminded of a story one of the mountain rescue team told me, of a group lost on nearby Lochnagar, at night, looking desperately for the torchlight of help, and wandering towards the lighthouse at Peterhead, sixty miles away.

The highest, the largest, the oldest, these are all human distinctions. The trees themselves don't seem to worry. And yet these three trees are pushing at the limits of what we know that trees can do and reminding us that nature has the capacity to surprise us. This scrappy rowan is the extreme high-water mark of the great revival of the landscape that is currently underway below us.

It's a long way back from here, and I decide to take the direct route over Derry Cairngorm. It's a bit of a trek over to the top, but I'm joined by the occasional ptarmigan, dapper grey with red eyebrows, croaking away. A mountain hare is cooling off in a snow patch, conspicuous in its blue-brown summer wear. There's moss campion and trailing azalea and three-leaved rush at my feet, plants which thrive in the cold and snow. A final twenty minutes to pick my way through a boulder field that is always more taxing than I remember, and I'm at the summit, 1,155 metres up.

From the summit I can see the whole estate, and much more besides. The Caledonian pinewood stretches out to the south and east. You can pick out a green fuzz across the open heather: hundreds of hectares of young trees, the next generation. Some parts are thick with growth, others less so. This is good – much of the ground is entirely treeless, and will remain so owing to quirks of ground conditions. To the east sits the Beinn a' Bhùird plateau, hunching over Glen Quoich. It is a huge hulk of a mountain, and yet it is only the eroded stump of a hill that once reached as high as the Himalayas. Glen Quoich is home to its own pine and birchwoods, and the Quoich Water, an untamed, braided river that feeds into the Dee at the floodplain of the Quoich Flats, which are something of a haven for long-legged, long-billed wading birds.

To the south is the River Dee. You could see its source from here but for the mass of Ben Macdui. Further south still sits Carn Bhac, a handsome hill and the southern edge of the estate, ten miles away. Following the Dee, we see more of the pinewood, through the gorge at the Linn of Dee, and out into the wetter, more open landscape of the west. Here sit the remote Munros An Sgarsoch and Carn an Fhidhleir, lumpy bumpy things, twelve miles away at the western limit of the estate, a vast expanse of moorland and bog, home to large heath butterflies, golden plover and short-eared owls. This is the Geldie beat, proper sporting country, a place of stags, grouse and salmon, of long wet days and peat-stained trews. North of the Geldie burn slouches the mass of the Cairngorms proper: Beinn Bhrotain, Monadh Mòr, Bod an Deamhain (The Devil's Point), Cairn Toul, Sgor an Lochain Uaine, Braeriach, Carn a' Mhaim, the names a liturgy for the Scottish adventurer, their distance and height leg-aching to look at. Dominating the foreground is Ben Macdui. The cliff face of Sputain Dearg is still covered in snow patches, and Lochain Uaine remains shrouded in ice. A raven cronks overhead, seeking me out for sandwich crumbs.

Further out, beyond the march, Beinn a' Ghlo, the misty mountain of the wilds of Atholl, sits to the south-west; Cairn Gorm is to the north, its weather station obvious. Further round: Beinn Mheadhoin; Bynack More, the edge of the Cairngorms; Ben Avon, with its tors; Lochnagar and the Mounth hills to the south-east; Morrone, Braemar's hill, of the foot race fame; Glen Ey and the expanse of the Grampians. Further afield, even: the Monadhliath, the Lawers group, and is that Schiehallion?

Derry Cairngorm sits in the heart of the largest contiguous stretch of land managed primarily for nature in the UK. To the north and west of the Mar Lodge march lies Cairngorms Connect, a partnership of private, public and charitable landholdings that comprises 600 square kilometres of land managed for wildlife, stretching from Glenfeshie to Abernethy. These are also sites of remarkable environmental regeneration, where the pinewoods are once again reclaiming the giant glens. Combine Mar Lodge and Cairngorms Connect and you get 900 square kilometres of land managed principally for wildlife, landscape-scale ecosystem restoration and the wider benefits that nature brings. It is a giant landscape of hope in the middle of Scotland.

Scotland is a nature-depleted place; the UK is ranked 189 out of 218 countries for the intactness of our biodiversity.[13] Over the centuries we have destroyed our island's wildlife and wild places, and the trend of extinction is increasing. Mar Lodge Estate retains only a part of what was once an incomprehensible ecological assemblage. But on a beautiful day in May, sitting on top of Derry Cairngorm, you'd struggle to notice.

Ecology is the study of the rules, processes and relationships that guide the proliferation of life on earth. It is rather beautiful.

Environmental conservation is, on the face of it, applied ecology, and should therefore also be beautiful. In actuality, environmental conservation is among the most complicated, underfunded, political, vitally important activities that a person can do. It is often rather ugly.

Environmental conservation is always an exercise in crisis management. For the conservationist working to save a species or a landscape, there is never a correct solution to a problem, only a least-bad solution. Conservation is a practice forged in the Anthropocene, the epoch in which humans have emerged as the primary agent of ecological and geological and climatological change on the earth. 'Preservation' and 'conservation' are poor words for what the anthropologist Thom van Dooren calls the 'practice of care that aims to nourish and sustain species and their living participants in far-from-ideal conditions, where the most desirable options simply are not available.'[14] Conservationists have always occupied philosophical and geographical edgelands, all too often forced to decide not which creatures get to live, but instead which creatures must be sacrificed so that others may survive. Perhaps because the movement has grown out of the margins, it has always been underfunded and often unappreciated. Sadly, it has become the norm that our national natural treasures are protected by a thin line of charities, subject to the whims and caprices of their members and the economy.

It is wrong to think that all conservationists see the world in the same way. Conservation is an evolving discipline. There is often fierce debate within conservation circles as to the best way to protect our environment. Many conservationists believe in their heart of hearts that the natural world is doomed, and with it will go not only that which sustains us and brings the greatest source of richness of human experience, but also the myriad ways of life that make up the infinitely complex tapestry of life on earth. From a human perspective, the true nature of this tragedy is that this great richness, the chance to interact with something infinitely more complex than the mere human, will pass by largely unnoticed.

And yet. The Mar Lodge experience, the complicated, messy business of conserving the creatures and plants and landscapes and cultures which forge what is best in humanity, shows us that hope for the future still exists. Humans and nature are adaptable and perhaps more resilient than we often think. Conservation work is hard, and underpaid, or not paid at all. It is often a case of fighting a rising tide. Yet for all its difficulties, environmental conservation can be a deeply holistic experience: as anyone who has planted a tree, or hauled a rock into place in a footpath, or stalked a deer can attest, it provides a link to the natural world which all of us crave, however unconscious that craving may be.

Conservation is changing. New crises and continuing losses to biodiversity are challenging the old ways of doing things. New words like 'rewilding' are entering the mainstream. A growing group of ecologists, anthropologists, green critics and activists is questioning long-held paradigms that have framed how we understand the environment for centuries. What do we mean when we say 'wilderness', 'wild land', or even 'wild animals'? How do we best go about saving the environment? In light of these questions, the anthropologist Hugo Reinert locates 'the wild' at a 'complex, awkward juncture in contemporary human–nonhuman relations: simultaneously an object of control and withdrawal, absence and intimacy, wildness and impurity; a site of complex and intractable controversies—but also, perhaps, of hope'.[15] It is this juncture that one sees at Mar Lodge: wildness and impurity, complex controversies, but also hope.

This book celebrates twenty-five years of Mar Lodge being under the protection of the National Trust for Scotland, a complicated, decidedly imperfect organisation, but one whose guiding aim is the protection of Scotland's heritage for the benefit of its people. This book is far from being the complete story of this land, its wildlife and its people, if such a book could ever be written. Instead, it is a collection of stories from the edge of environmental conservation. All these stories are 'edgy' in some way or another. Some use cutting-edge technology, genetic sampling, synthetic pheromones, satellite tracking. Others are politically edgy, pushing for new ways of visualising and interacting with the Scottish landscape. Some stories are of niche species and niche conservationists, at the edge of what most people would consider worth being interested in at all. Many of the species and habitats involved are on the extreme edge of their world distribution; still others are on the edge of extinction. We'll be exploring the highest, coldest mountains, the clearest waters, the remotest corners of Scotland. We are being kept company by some of Scotland's most exciting wildlife: golden eagles, rutting stags, salmon. We are also being kept company by people working at the top of their game, forging new and innovative ways of protecting and enhancing the natural world for the rest of us, and in some instances from the rest of us. We are venturing to the conceptual edges of conservation, posing questions, dissecting the assumptions and science and cultures that underpin some of the biggest questions of all: what is nature? What does it mean to conserve the environment, and how do we do it? Why bother?

This book is an account of the work of just some of the people who dedicate their lives to the creatures that others neglect, abuse and destroy. This book is about the myriad motivations that fuel their work: the love, the rage, the sheer sense of injustice which all people who would call themselves conservationists must eventually hit up against. It is conservation on the edge: dangerous, exciting, challenging, confusing, difficult, controversial, contradictory, experimental, fun.

I've split this book into three parts. From an ecological perspective, this is entirely arbitrary. From a human perspective, this is essential. For these three landscapes, and the challenges that face them, are bound to the cultural conventions that have arisen from their perception by humans. First, we'll be delving into the ancient woodlands, a fragment of an almost lost landscape. Here, we'll meet gamekeepers, bryologists, geneticists, climatologists, mosses, capercaillie, deer. We'll be looking at how the woods were saved and discussing what they may look like in the centuries hence. Next, on the moors, we'll be delving deep into one of the most politically controversial landscapes in Scotland. We'll be stalking deer, fishing for salmon, satellite-tagging rare birds of prey, tracking down rare wading birds, and seeing how we can help the landscape help us in the fight against climate change. We'll be looking at some of Scotland's most difficult conservation conundrums. We'll meet conservationists coming up against rampant illegality and wildlife persecution and see how the moors came to be loved and loathed in equal measure, and how our attitudes towards them need not be so polarised in the future. Finally, we'll be heading up into the mountains. We'll be finding out what drives people to visit this beautiful, dangerous place, how conservationists protect it from those very people, and what people are doing to save the creatures that call it home. We'll be rediscovering lost ecosystems, pondering the effects of climate change, and wondering what might be in the future.

In 1995 a new owner, working to keep the land in trust for the benefit of the nation, took on responsibility for this remarkable land. It would be a haven for the UK's rarest wildlife, with flourishing woodlands under a ceiling of mighty, subarctic mountains, open for everyone to enjoy. It would be a model sporting estate, where shooting and conservation worked hand in hand for the benefit of all, human and beast alike. What could possibly go wrong?

Part One
IN THE WOODS

For the connoisseur of British woodlands, the Cairngorms' Caledonian pinewoods take some beating.[1] They are ancient landscapes. The grand old trees, the carpets of juniper, heather and bright green blaeberry, the sounds and smells: the Caledonian pinewoods have their own quintessence. These are places where trees have grown uninterrupted through natural processes for 8,000 years. In so doing, they have adapted to the land and have become genetically distinct from Scots pine in other parts of the world. Of all the iconic landscapes of Mar Lodge, it is the Caledonian pinewoods which are the most celebrated. They are home to an incomprehensible assortment of creatures: crossbills, red squirrels, pine martens, black grouse, capercaillie, siskins and bramblings, spotted flycatchers, redstarts, tree pipits, goshawks, owls, ospreys, golden eagles, red and roe deer; rare plants: twinflower, orchids, wintergreens, globeflower; and invertebrates: wood ants, pearl-bordered fritillaries, longhorn beetles. Then there are the birches, the willows, the junipers, the aspens, the rowans. They have an authenticity, a presence, a continuity. 'To walk through them is to feel the past,' wrote H. M. Steven, the pioneer of Caledonian pinewood conservation, back in the 1950s.[2]

And yet there are two Caledonian pinewoods: the pinewood of history, myth and culture, and the pinewood that exists in the glens. The two have a very difficult relationship. To talk of Caledonian pinewoods is to evoke the ghost of a lost great forest, swarming with bears, wolves and Picts, which spread from coast to coast in an impenetrable wall from which even the mighty Roman legions recoiled. It is a habitat which evokes national pride, harking back to an earthier, wilder nation of great power and untamed spirit. Since its first coining by those very Romans who failed to conquer Scotland, the Caledonian pinewood has suffered from the overbearing weight of mythology. It has been othered, made into a wild place, an uncivilised landscape, a place to visit rather than linger, where the visitor may reconnect with their lost origins, but only for a holiday. This does a great disservice to the people who have made their homes and living among the pinewoods for thousands of years, and indeed, to the history of our forests. The truth of our relationship with the woods of Mar Lodge is much more gloriously complicated than the myth of the great wood of Caledon.

Nevertheless, such mythologies are potent, and valuable in their own right. They call out to the wild heart lurking within us all. They tell us how we want to think about ourselves, and what is missing from our lives. Complications of history and myth aside, the remnants of the Caledonian pinewood are truly a last, vital link to a 'deep time', one which stretches beyond the surface world of the postmodern. Wildcats may not wait behind every tree, but then again, they just might. To walk through the woods is to feel the past, but also to be haunted by the ghosts of what has been lost, and the portents of what might yet be.

The challenge for conservationists, then, is to unpick the truth from the myth, to navigate the thorny questions about what consists of the reality of the Caledonian pinewood and what does not. Step forward Paul Ross, dendrochronologist and climatologist from St Andrew's University, who is here to get some samples for his PhD thesis. Paul is part of a team that has been studying the history of Scotland's Caledonian pinewoods for the last decade and more. I join the team, Paul, Rob Wilson and Emily Reid, for a day of sampling in the Dubh Ghleann, a particularly remote and beautiful remnant of the pinewood at the top of Glen Quoich on a cold March day.

As I am driving us up the glen, Paul tells me what the team's up to. Actually, he doesn't, at least for a while, because Paul's an ecologist both by training and by inclination and we keep seeing wildlife that excites us both. Behind us, Rob and Emily are getting impatient with our conversation, so Rob jumps in. 'We take tree core samples from old pine stands across Scotland. These give us all sorts of information about the woods, beyond how old the individual trees are. Ultimately, we want records of all the surviving pinewoods in Scotland, and we're nearly there. We're working further afield now too. I have just got back from Argentina, and have a similar project going on with monkey puzzle trees. Actually, Emily's currently looking to set up another, similar project in the Yukon.'

Rob's been here a lot since 2006,[3] when the project started. It's thanks to him that we know how old the Old Tree is. He also runs the ridiculous Lairig Ghru race, a marathon and more through the mountain pass that cleaves the settlements north and south of the Cairngorms massif. 'I'll keep doing it until I get a time that I'm happy with,' he says, which perhaps explains the remarkable output of his team.[4] He points out the jungle of newly regenerating young trees that we are travelling through. 'I don't remember all this,' he says, 'where's all this come from?'

We get to work, eventually. The Dubh Ghleann is at the furthest extent of the pinewood, and we are walking right to the treeline. Here, Scots pine is getting close to the edge of what it can cope with climatically. We are working 550 metres above sea level. The team works meticulously, with the skill and jocularity that come with competence. (One of life's great joys is working with competent people.) As with most ecological sampling equipment, their kit is both extremely clever and extremely spartan, designed to cope with the demands of decades of underfunding. The tree corer is a fantastic piece of Swedish design, unchanged for more than a century. It has the gravitas of kit that is both functionally perfect and aesthetically pleasing, like crampons and Trangia stoves. It is essentially a small but long, hollow screw with a handle and a long insert, which extracts a tiny core from the tree. This allows you to determine the tree's age, and much more besides, without harming the tree in any way. These cores are then stored in large drinking straws, labelled and capped with masking tape. This storage solution, despite being extremely homemade, is apparently infallible. Emily is pleased to be back working with pines. Monkey puzzles, she says, are horrible to work with. But pines give excellent cores.

Rob shows me a core. It's a beautiful thing, a long cylinder of tiny bands of alternating light and dark wood. But more than that, it's a record of the age and growth rate of the tree. More even than that, by comparing the different amounts of 'late summer growth' and cross-referencing them with known summer temperatures from recent decades, the team can work out the summer temperatures from centuries ago. It's a beautifully neat example of ecological and climatic interactions. The pinewoods are a living record of centuries of summer weather.

Rob is a climatologist by trade, and this was actually how the whole project started. Rob realised that by collating and referencing the growth rates of trees across Scotland he could draw a picture in time of average summer temperatures going back, in theory, for as long as there have been trees in Scotland. He was originally drawn to pine for its longevity in the landscape. A study like this can't work with other Highland tree species like willow, birch and aspen, which do not grow to a ripe age and also rot quickly. 'I love working in the Cairngorms – the trees here behave themselves. There's less noise in the data than the west-coast pinewoods.'

Noise?

'Disturbances in the tree-ring data. One-off events like felling large areas of trees can affect the growth rates of surviving trees, which produces inconsistencies in the data.' The problem, says Rob, is more significant in the West Highlands. 'I've not come across a single pinewood in the west which is behaving itself. It has warmed up in recent years, so they should be growing more. In fact, we're seeing the opposite. Why is that?'

He corrects me on my coring technique – I'm too wobbly, and keep breaking off the most recent decades of growth. 'We've got a wobbler here! Paul won't be impressed. Those most recent rings are the important ones for his work!'

In fact, it is the noise – the inconsistency in growth rates between trees in the east and the west – which is the focus of Paul's work. 'My research is on the ecological relationship between trees, climate and the ground flora around the trees.' Paul is indeed unimpressed by my coring technique. He's even less impressed when, carried away with conversation, I core a dead tree instead of a live one.

'I don't think we should tell anyone about this,' I say.

'Oh yes, I think the whole team needs to know!'

For Rob, the ageing of individual trees is a by-product of generating climate information. But it is a by-product that has greatly improved our knowledge of the recent and not-so-recent history of the pinewoods and has provided a huge amount of information on ecological hypotheses that were once untestable. 'At first all we were interested in was the old trees, but now they are less important to our work.' What gets Rob really excited now is sub-fossil pine, 'the stuff you find preserved at the bottom of lochs'. This stuff can be hundreds or even thousands of years old, so it pushes the information back further in time than the live trees we are working with today. His plan is to create a pine-based chronology of summer temperatures going back through the Holocene, the time after the most recent Ice Age. 'We have material going back eight thousand years, although it will likely not be possible to create a continuous record back to that point. Two thousand years is theoretically possible, however. We already go back to the tenth century.' He's talking me through it – how they're looking for pine beams in old buildings, and searching out new lochans to go scuba diving in for pine stumps – when I notice a red kite, a rarity here, and start shouting at Paul to inform him of the fact. Paul's suitably impressed, but Rob is obviously less pleased to have an avian distraction. 'Bloody ecologists,' he mutters, and I'm only fairly sure that he's joking.

So what can the team tell us about the Caledonian woods? And what of Mar Lodge?

Rob is keen to sort the facts from the myth. 'It's a lot more complicated than some people think.' The idea that there was once a continuous wall of pines stretching from coast to coast across the Highlands is wrong. First of all, the Caledonian woodland wasn't all pines, it was a mixture of tree species, depending on local soil, weather, altitude, aspect and the local population of herbivores. Instead of a great, dark, impenetrable wall of trees, ecologists now imagine a more diverse and indeed biodiverse landscape, of wooded glens and open hill tops, where trees naturally graded into bog, moor and tundra. This was a rich tapestry of broadleaves and pine, where the interactions of woody plants, grazing animals and human and animal hunters created a web of open glades, dense thickets and mature woodlands. This complex history puts the lie to our notions of woodlands as being landscapes with trees and moorlands as landscapes without trees. The natural ecology of Scotland is much more complicated than that.

This is not to say that Scotland was not once much more wooded than it currently is. The greatest extent of the pinewoods, and indeed woodland cover in Scotland, was around 6,000–7,000 years ago.[5] Since then, the woods have retreated through the twinned impacts of human actions and historic climate change, with human activity becoming by far the greatest cause of deforestation in the last two millennia. In fact, the lowest extent of woodland in Scotland occurred around the late eighteenth century, when woodland cover was as low as 4 per cent of the landmass (it's now hovering around 18.5 per cent, though much of this is non-native, often environmentally unfriendly spruce plantations, and it remains well below the 38 per cent European average).[6]

The oldest record of pines in Deeside is from around 8,000 years ago, roughly the same period as the arrival of humans to the area (both of these records are from Mar Lodge ground).[7][8] There certainly used to be much more pinewood at Mar Lodge. Go back 2,000 years and there was a pinewood in Glen Geldie, which is now a rather nice blanket bog, where ancient pine stumps emerge out of the peat. Historic climate change has a part to play in the story of the retreat of the Mar Lodge woodlands. But from the Middle Ages onwards, and particularly post seventeenth century, the Mar Lodge pinewoods have been subjected to a long war of attrition at the hands of humans. The woods were a source of fuel for local crofters and of timber for the navy.[9] Glen Geusachan, the little pinewood glen, which now hosts a mere handful of birches and rowans, probably held a remnant of Caledonian pinewood until the early modern era. A map of the estate from 1703 shows no trees in Glen Geusachan so we can be fairly sure that by this point there was, at most, an unharvestable number of trees left in the glen. But the fate of the Glen Geusachan pinewoods remains uncertain.[10] A number of sawmills were in operation on the estate in the eighteenth century, the ruins of which can still be seen in the Derry and the Quoich. The first record of sawmills on the estate comes from as far back as 1695, and we can be assured that the woods had been used as a timber resource for centuries beforehand.

But so long as the forest can regenerate, with young trees replacing old, forestry operations are not enough to destroy a pinewood. Indeed, the longevity of the Cairngorms' pinewoods points as much to the sensitive use of the pinewood resource by centuries of local inhabitants as it does to the difficulty of extracting the wood from remote glens. Paul reminds me of the startling fact that the pinewoods 'move'. New seedlings most readily grow in light areas, outside of the shade of the current extent of the woodland. Over the generations, this means that the trees move about in the glens. Back in the eighteenth century, this fact was observed in legal documents pertaining to the Mar Lodge and neighbouring woodlands – at this point, grazing was not halting regeneration, though there is evidence of foresters being annoyed by local cattle grazing on newly emerging pines.[11] The pines have ebbed and flowed across the country in waves, as climate and grazing pressure has changed. Sometimes they will have been able to survive higher up the glens; other times their extent will have retreated. Now, we are seeing pines racing up the hills, growing far higher than people thought was possible.

According to the wandering Reverend Charles Cordiner, who toured the Highlands in the eighteenth century, by 1776 a large part of the woodlands in Luibeg and Glen Derry had been felled, but there was also young regeneration coming up in Glen Quoich.[12] Rob shows me a core. 'I'm an impatient man. What I like about tree cores is you can see what's going on immediately.' His expert eye points out that this tree is about 250 years old. This is one of the young trees seen by Cordiner all those years ago.

In fact, the team's findings fit perfectly with surviving historical records for the whole of the estate. Since modern times, large-scale felling at Mar Lodge seems to have occurred across three main periods: the mid eighteenth century, when new technology combined with a new, extractive view of the land that allowed the forests to be exploited more efficiently than before; 1810–60, during the zenith of the Royal Navy's dominance of the oceans; and the Second World War, when the army needed all of the timber it could lay its hands on.[13] Most of the pines in Glen Derry probably grew up after the felling which was done before Cordiner's visit in 1776,[*] while the trees in Glen Quoich are those that he saw as young seedlings at that time.

By the early nineteenth century, however, the forests were depleted. This did not go unnoticed, nor unchallenged, as Elizabeth Taylor[†] attests, on viewing the recently deforested Linn of Dee and Glen Lui in 1869:

> What a noble spectacle this valley must have been in the height of its woody glory! How it would enhance the grandeur of these mountains, when their rugged slopes and

---

[*] The Old Tree, sitting up on the scree slope, probably got a stay of execution at this point because it wasn't of any use for its timber.

[†] No, not that one.

precipitous sides were hung with one continuous sheathing of fragrance and verdure! . . . It is stated as fact, that in five years' time 80,000 of these hoary veterans of the forest were cut down in this part of the valley.[14]

This is a good example of a newly emerging sentiment that had been growing out from the roots of the Romantic movement. Up until the eighteenth century, or thereabouts, there was a tendency among Western civilisations to treat nature, particularly wild and ferocious nature, as something to be feared, fought against and ultimately subjugated. This is the culture from which the myth of the Great Wood of Caledon emerged, all those centuries ago, when the term was first coined by the Romans. Taylor's sentiment is interesting because it expresses a fetishisation of nature and savagery, and lament for its loss for rude economic gain. The loss of the woods is to be lamented because woodlands 'enhance the grandeur' of the mountains. This sentiment is the flipside of that which led to the denigration of nature in earlier centuries. Either way, nature was perceived quite explicitly as something 'out there', set apart – the opposite of civilisation.

In such feelings, then, we can see the beginnings of what the historian Chris Smout calls the 'roots of green consciousness'.[15] For the first time, people were coming to value Mar Lodge's woodlands for their 'nature'. It was specifically their wildness – the fact that they were 'uncivilised', that gave them an allure for a new type of rich, well-connected visitor. But as we have seen, Mar Lodge has been a home for humans for millennia.

Here we must pause to consider the long-reaching, complicated influence of Queen Victoria on the Mar Lodge woodlands in more detail. Her Majesty was famously a great lover of the Highlands, and of Deeside in particular. She lamented the loss of the Glen Lui woods herself: 'Re-entered our carriage & proceeded back on the other side of the Dee, through Glen Lui, once full of the finest fir trees but of which only the stumps now remain, Lord Fife's creditor having cut them all down.'[‡][16] So upset was the queen at the thought of the imminent felling of the glorious Caledonian pinewood of nearby Ballochbuie that she bought the woods herself. Deeside would in all probability have lost even more of its woodland than it has done, were it not for royal influence.

And yet the influence of the 'Balmorality', the royals and their hangers-on, would go on to do huge damage to an already depleted landscape and culture. Queen Victoria was drawn to Upper Deeside by the deeply attractive idea that the landscape was an empty wilderness – a blank, wild canvas. The landscape that the Balmorality created was a confection – an ersatz wild landscape – and it was created on the back of an oppressed, depopulated landscape. The idea that the Highlands were some untamed wilderness, its only inhabitants 'noble savages', was a direct result of centuries of oppression of the Highlands which culminated in Culloden and the Clearances.

The Victorians were both the pinewoods' conservators and destroyers. Where the landscape did not fulfil its wild aesthetic promise, steps were taken to beautify the landscape. The Victorians were 'reforesters' in their own way. They planted up the grounds around their newly erected hunting lodges, and replanted areas of woodlands lost in previous rounds of felling. But it is fair to say that they were less interested in the ecology of the landscape than its aesthetic value. The woods around Mar Lodge and Derry Lodge, planted at this time, are filled with non-native firs and spruces. European larch was particularly favoured for its rich autumn colours. Worse, the deer that they hunted, and their management of the moors for red grouse, stifled native woodland regeneration for well over a century. The rise of the red deer meant the loss of both people and pines from the glens. By the 1950s, the ecologists H. M. Steven and A. Carlisle catalogued all of the remaining Caledonian pinewoods in Scotland, and discovered that they had

---

‡ What goes unremarked is concern for the fate of the crofters who had been cleared from the glen at roughly the same time as the pines.

been reduced to just 16,000 hectares, maybe 1 per cent of their historic area, of which around 815 hectares could be found at Mar Lodge.[17] Some of these pinewoods were mere scraps – just a few trees scattered loosely across an otherwise treeless glen. In other places, including Mar Lodge, the woods were more substantial. But in almost all places, the woodlands covered little more than a shadow of their historic range. At Mar Lodge, by 1995, the woodlands were on their knees.

The samples taken, we retreat from the Dubh Ghleann. Back at the lab, Paul will fix the cores in wooden mounts, sand them with extremely fine paper, and finally take an extremely high-resolution photograph of them. These images will then be run through custom software, designed to glean every scrap of information that it can from the cores. This will be fed back into a database to inform Paul's and others' work. The science of researchers at St Andrew's is fleshing out our understanding of Scotland landscapes, giving us new, detailed insights into our woodlands, what they used to be like and what they could be like again. They are allowing us to see through the fog of myth and history, giving us a glimpse of the unique individual history of each tree and each patch of woodland. Their work has greatly improved our understanding of Mar Lodge's woodlands, put them in their national context, and allowed us to understand their value in a way that was previously impossible. It sits alongside the work of geographers, ecologists and cultural historians. To walk through the pinewoods is to feel the past. But it is also to imagine a new, better future for both humans and nature in these glens – the regeneration of an environmentally sensitive woodland culture.

There remains a lot of information that has been lost, and a lot of unanswered questions. What of the broadleaf trees, the birches, rowans and willows, which tread more lightly on the landscape and do not leave a mark like long-lived, slow-rotting pine? Did Beinn a' Chaorainn, the hill of the rowan, once harbour a broadleaf woodland on its slopes? What of Allt a' Chaorainn, the rowan burn, on the other side of the estate? Or Clais Fhearnaig, the hollow of the alder, in which a single elderly alder can be found, now joined by its regenerating progeny?

The balance between what is known and what we need to find out about the woods is what is informing the work of the conservationists in this first section. We'll be looking at how a team of ecologists and stalkers kick-started the regeneration of the woodlands for the first time in centuries, and how that venture turned into a decades-long labour of love and battle of wills that continues to this day. We'll be paying close attention to moss, joining bryologists as they search for rare species among the deadwood. We'll be getting up very early in the morning to look for woodland grouse, and staying out very late in the evening in search of roe deer. We'll be luring moths with fancy pheromones, finding out what really goes on in an ant nest, and meeting the new human inhabitants of the woods. Onwards!

1

Scots Pine

> Nothing in the world is worth having or worth doing unless it means effort, pain, difficulty.
>
> Theodore Roosevelt, creator of the US national parks, *American Ideals*, 1897

'Scots pine, 58.7 cm, 13.5 cm, 75–100, single stem, 3+5 new. Scots pine, 27.5 cm, 25.3 cm, 35, multi-stem, no browsing, code is 0.' It's August, and Shaila is measuring seedlings. Hunched over, her favourite old-school folding wooden ruler in hand, head tucked deeply into a midge net, Shaila records the height of trees, which shoots have been browsed, the number of shoots. She works and talks without stopping, measuring hundreds of seedlings in a sitting, barely pausing to catch a breath. She knows the trees well by now; she has measured some of them for seventeen years, and recognises them from year to year. Occasionally, a particularly interesting tree will get a comment, a 'wow, it's a monster' or, a 'sorry specimen'. I'm on the computer, an old, beaten-up Toughbook, inputting the data, happy that the breeze is keeping away the worst of the midges, only slightly put out by the drizzle. A couple of days ago, the corpses of thousands of midges mixed with the rain, creating a soapy paste that threatened to leach into the interior workings of the computer. It is meticulous work, but Shaila works quickly, and it takes a lot of concentration to keep up with her. Sorrel, Shaila's collie, sits nearby, contemplating whatever dogs contemplate while humans play with trees. After eighteen years of monitoring, Shaila has around 60,000 tree-height measurements, and hundreds of thousands of individual pieces of data.

Dr Shaila Rao is the estate ecologist: the in-house go-to person for all things ecological. It's her job to provide the science that informs management decisions on the ground. Between you and me, Shaila has been the driving force behind some of the more remarkable recent changes to the Mar Lodge woodlands. She's been here since 2002, an immovable object against the Trust's endless cycle of restructuring and redundancies. Her doctorate was in mountain hare ecology, specifically the relationship between grazing animals and woodland regeneration. Driven, clever, focused, she's the sort of person who has a favourite ruler, but you couldn't hold that against her. She wandered for a few years, as ecologists often do, putting time in working with arctic foxes in Lapland, sandhill cranes in Mississippi, and trekking through Alaska, before settling into the challenge of saving the Mar Lodge woods. Since then, among many other things, she has been counting and measuring tree seedlings in seventeen quadrats (10 metre x 10 metre squares) strewn across the woods. Some quadrats are set within the mature trees in the heart of the woods, others are at their edge. Some are high up, at the furthest extent of pine growth, some are in small groves of birch and aspen. She measures the trees and records what's been eating them. The plots are scattered across three glens and a hundred square kilometres. There are big walks involved to get to some of them. The work takes two people two weeks. It is wet, midgy, awkward, uncomfortable, back-breaking. Today's quadrat is in Glen Derry, on the edge of the pinewood, with a frankly phenomenal view of the pines stretching out towards Derry Cairngorm and Beinn Mheadhoin.

In 1995, the Trust embarked on its plan to save the pinewood. It was decided that the number of deer would have to be drastically reduced to give the woods a chance to regenerate. This plan would be in harmony with the wider sporting interests of the estate. Grand monitoring schemes were drawn up. Deer were stalked and sent off to the butchers. Environmentalists eagerly awaited the revival of the pinewood. Sporting interests also watched on, some curious, some excited, some concerned that the grand scheme would bring about the end of their way of life, and their income.

Then, for over a decade, no new trees grew. Nothing happened.

To understand the Highlands, you need to understand two species above all others: the Scots pine and the red deer. Both species have existed on Mar Lodge for the last 8,000 years. Both species rubbed along together perfectly well for 7,000 of those years.

The Scots pine is Scotland's favourite tree. It is a keystone species, which underpins the whole ecology of the Caledonian pinewood. They are tough, creating forests in conditions that other trees struggle with, bringing with them hundreds of species which enjoy living in and amongst pine, including red deer. For red deer are, at heart, woodland animals. They like being in woodlands. And woodlands like red deer: their grazing opens up areas of dense growth, allowing other species to flourish. Their trampling creates bare patches of earth where seedlings can germinate. Their dung recycles nutrients. They are prey for wolves, their carcasses provide food and a home for their own assemblage of species. Red deer grow bigger and stronger in woodlands than on the open moor. Red deer are brilliant animals: resilient, adaptable, clever, beautiful. They reproduce quickly when conditions are good and when there are no predators. They regularly top polls of the nation's favourite animal. People love deer, and why wouldn't they?

Deer are also wild animals, and in a country generally lacking in big native wildlife, they are striking for their size and wildness. But there is a close cultural relationship between humans and deer, closer indeed than their wildness suggests. Deer stalking has been a tradition for centuries. Many estates feed their deer in winter. Deer may be wild, and therefore not owned by anyone, but the landowner has responsibility for their numbers on their land and owns the right to hunt them. Stag hunting has grown to be the major land use in the Highlands. For this reason, in the Highlands, land value is closely tied to deer numbers: the more stags you have, the more valuable your land.

Pine and deer belong together. The problem is, there are too many deer, and that's because of humans. A lack of natural predators, combined with a culture which has venerated them in both life and death for centuries, has fostered the perfect environment for a deer population boom. A study of Europe's red deer from the early 2000s estimated that there were 355,500 red deer in the UK, of which the vast majority (300,000+) were to be found in Scotland.[1] This means that there was roughly the same amount of red deer in Scotland as in Germany, France and Italy combined. The only country that came close to having a similar number was Spain, a country six times larger than Scotland. Current estimates from Scottish government put the red deer population at 360,000–400,000. This means that, in spite of heavy culling in nature reserves across the country, and in spite of widespread acknowledgement of the 'deer problem' at the highest levels, the number of deer is not decreasing.

Some grazing is vital for a woodland ecosystem, but too much grazing will cause it to disappear. When the pines disappear, so too do the myriad ways of life that comprise a complex woodland ecosystem; the wood ants, the plants, the mosses, the birds. It has been proven again and again across Scotland that if you reduce deer density to numbers more regularly found in Europe then you gain in overall biodiversity.[2] [3] [4] [5]

The famous Cairngorms naturalist Seton Gordon was worrying about the state of the Mar pinewoods at the turn of the twentieth century. In the 1940s, over in the US, the hunter/ecologist Aldo Leopold described the tragedy of over-grazed, over-deered mountains, which looked as though someone had given God new pruning shears and forbidden him all other activity.[6] In the 1950s Frank Fraser Darling, who studied the deer of An Teallach on the west coast of Scotland closer than anyone had studied deer in Scotland before, complained that the scourge of red deer was stifling woodland growth, reducing species diversity, creating a 'wet desert'.[7] There were around 150,000 deer in Scotland back then, less than half the current number.

Influential though her work may be, Shaila is just the latest in a long, long line of people monitoring the health of the Mar Lodge woods, and the relationship between these woods and their deer. The progenitor of Shaila's monitoring programme was set up in the 1950s, when part of Mar Lodge Estate was included in the NCC's[*] Cairngorm National Nature Reserve. The reserve warden Malcolm Douglas set up a very basic and very effective experiment. He erected small 'exclosures', fenced areas of pinewood into which deer could not enter (hence, they were 'excluded' rather than the trees being 'enclosed'). Then he sat back and saw what happened. And what happened was that young trees grew again. In 1959 Douglas went bigger and stuck a fence around 2.5 hectares of wood in Glen Derry. The results can be seen today – a square of sixty-year-old trees, surrounded by older granny pines and the young regeneration that has sprung up in the last decade. Sixty years ago, Douglas showed that the pinewoods were not regenerating because of overgrazing by deer, and that if nothing was done then sooner or later the pinewoods would be lost.[8]

Shaila's quadrats are the direct descendants of Malcolm Douglas's experiments, and the experiments of many others across the intervening years. But what is different is the scale. Where others were forced to play with fences, under Trust ownership conservationists could experiment with a landscape-scale solution to a landscape-scale problem.

I mention fencing in a lull between taking measurements. Why not just take Douglas's experiments to the extreme, and stick a fence around the woods? It's clearly a question she has heard a lot. 'It treats the symptoms, not the cause, and causes lots more symptoms along the way. Fencing off different pockets of land for different uses is the opposite of an ecological mindset. We need to think of landscapes

---

[*] Nature Conservation Committee, the first in a long line of endlessly changing government bodies charged with protecting the nation's nature. It is the ancestor of the agency-formerly-known-as-Scottish-Natural-Heritage, now known, of course, as NatureScot.

holistically.' From the outset the Trust decided not to resort to erecting deer fencing where possible. Deer fencing can be very good at helping trees to grow. Deer fencing is also expensive, and fails a lot, particularly in snowy places. It kills a lot of wildlife, including capercaillie and black grouse, which collide with fencing while flying fast and low through the trees. A single deer fence installed in the early 1990s, just before the Trust took ownership, was responsible for several dead capercaillie.[9] Fencing disrupts wildlife movement and ecological processes. It also looks horrible. It gets in the way and reduces the wild character of the place.

Nevertheless, landscape-scale deer population reduction was a controversial decision. At the time, with one or two notable exceptions, restoring woodland without fencing was not the done thing in Scotland. It would be the first time that such an action had been taken in Deeside, an area famous for deer stalking, grouse shooting, salmon fishing, and the royal family. It may have been a logical decision, but nevertheless it was an extremely bold decision for the Trust to take.

From the outset it was also realised that, at least until the woodlands were regenerating well, the estate would have to be split into two zones. These were called the regeneration zone and the moorland zone. The moorland zone, the area to the west and south of the Dee, was to be managed as a deer forest and low-intervention walked-up grouse moor, as per the agreement with the Easter Trust. Here, deer numbers would be managed to conserve the moorland habitats. The plan was to reduce the number of deer in the regeneration zone from 1,500, and upwards of 2,000 on occasion, to a resident herd of 350. This, it was thought, would give the beleaguered pines the chance to grow new seedlings and for the woodland to expand into places where it had been lost.[10] The total number of deer on the estate would be reduced from around 3,500 (with a high of 5,500 in 1993) to 1,650.† This work would be undertaken by the Mar Lodge game-

---

† The high-water mark for the Mar deer herd was the mid 1800s, when as many as 8,000 deer were present on the estate. That the land could hold

keepers, alongside their usual tasks of taking clients out hunting deer and shooting grouse on the open moor. Simple.

Such was, and remains, the controversial nature of deer culling that significant funds were put into monitoring efforts. The Trust needed accountability. If things weren't working, they would need to know why, and quickly. There was another, legal, edge to this work. The woodlands of Mar Lodge are a priority habitat within the Cairngorms Special Area of Conservation – a European designation which protects our rarest habitats. If they are not in a healthy condition, or at least returning to health, then the landowner is effectively breaking the law.

It quickly became clear that landscape-scale deer reduction was not a simple task. In 2002, Shaila took on responsibility for all the ecological monitoring on the estate, and it was immediately apparent that after seven years of controversial deer management there was little sign of the woodland recovering. That first year, the average size of the seedlings in her quadrats was 9 centimetres.[11] There were seedlings in amongst the heather, but as soon as they grew above the height of nearby vegetation they were chomped back again.

Shaila watched this happen for five years. 'It was desperate. It was totally frustrating for me in the early years because you were setting up monitoring, providing information, showing that the majority of seedlings were being quite heavily damaged, but we weren't seeming to respond to that in terms of management. In the quadrats it could take an hour just to count the deer dung. I used to look up and see a group of deer sitting above me, watching me work.'

Back in 1998, to complement annual surveys, 3.7 hectares of woodland were monitored in extremely close detail. This would be repeated once a decade. All the mature pines and birches were measured and mapped. The height and species composition of the vegetation was recorded, as were wood ant nests. All of the deadwood,

---

as many as that tells you just how much those very deer depleted the ability of the land to support wildlife in the intervening years.

the fallen branches and tree stumps, was measured. Fixed-point photographs were taken. In 1998, just four seedlings were recorded above the height of the adjacent vegetation in these 3.7 hectares. In 2008, the number of seedlings had increased to just 44. The survey also found some mortality in mature trees: pines were dying at a rate of 2.3 per cent per decade, a number which would increase as the effects of ageing (senescence) kicked in. Worse, as trees get older they produce fewer seeds. The older the forest, the harder it is to regenerate. The birches, being shorter-lived trees, were being lost at an alarming rate of around 10 per cent per decade, and not a single young birch had grown up to replace them.[12] By 2008, Shaila's research showed that despite management efforts, and despite some evidence of seedling growth, deer were still suppressing regeneration. That year the average size of her trees was 14 centimetres. At the rate of growth recorded, the pinewoods could well be lost. 'It's not rocket science after all. There were times when I thought, oh God, if we don't do something soon we'll lose the woods altogether.'

It certainly isn't rocket science. Across the march, another estate was about to spectacularly prove this.

Glenfeshie is among the most beautiful glens in the Cairngorms; 'the jewel in the crown', as it is often known. It is the western boundary of the Cairngorms massif, and shares a march with Mar Lodge along Glen Geldie.

It has shared a remarkably similar ecological history to Mar Lodge. The 17,000 hectare estate harboured significant remnants of Caledonian pinewoods but had been ravaged by extremely high densities of deer for well over a century. As at Mar Lodge, statutory bodies became extremely concerned by the dying woodlands. Things got so bad that, in January 2004, the Deer Commission took matters into its own hands, using contract stalkers to significantly reduce the numbers of deer in the glen. This action caused far-reaching uproar in sporting communities across Scotland. During a particularly heated debate in parliament in 2007, Fergus Ewing MSP, who currently serves as cabinet secretary for rural economy, said, 'One is reminded of the slaughter of deer at Glenfeshie not so very long ago, which stimulated an outrage ... the idea that that animal, uniquely, should be subject to state slaughter seems to be at odds with the approach to animal welfare issues that is adopted with every other species.'

Next, the Liberal Democrat MSP John Farquhar Munro raged against the 'indiscriminate and disgraceful mass slaughter of red deer at Glenfeshie'.

In 2005, the estate was purchased by Danish billionaire Anders Povlsen. Unencumbered by tradition, the estate went hard on their deer, continuing where the Deer Commission had left off. Ecologically speaking, despite the remonstrations of certain parliamentarians, it was wildly successful. By 2012, Glenfeshie would prove that spectacular pinewood regeneration could be achieved across huge areas of land, simply by reducing the number of deer in the glens. This controversial management approach was nothing short of a revolution. The monarch of the glen had been dethroned.

After a decade of controversial management and little to show for it, the period from 2008 to 2012 was a difficult one at Mar Lodge. Ecologically, things had been so bad for so long that a concerted effort to clear the woods of all but a token presence of deer was needed to give the trees the fighting chance they needed. The Mar Lodge management team decided to double down on their strategy. Extra deerstalkers were brought in to reinforce the gamekeepers on the ground. Helicopters were used to spot the deer, and to transport stalkers into the hills. Scottish Natural Heritage gave the Trust a licence to shoot deer at night, and outside of the sporting season. As deer numbers were reduced, the lack of competition and improving food source created a vacuum that sucked in deer from neighbouring estates. Nevertheless, the deer density in the regeneration zone went below three per square kilometre. Subsequent research would show that this density was the magic number at which regeneration would occur at Mar Lodge, for the first time in centuries.[13] ‡

But the work exacted a heavy toll. 'It was a stormy period of time,' says Shaila. 'When I look back, it was awful. There was a kickback from some of the neighbouring estates, and internally there was mixed feeling about the policy. Some people felt it was futile that we were shooting all these deer when there was nothing to show for it. And part of it was about the manner in which we were killing deer, and not putting up fencing. We had two or three meetings in the village where you could have cut the air with a knife. It was very unpleasant for the people working here.'

In 2010 the head stalker quit, two years before retirement, explaining that he considered deer to be an endangered species on Mar Lodge Estate. It was a blow. He was a fixture at Mar Lodge, well respected in the local community and the sporting industry, with decades of experience. The Scottish Gamekeepers Association was particularly hostile to the management at Mar Lodge, as they had been with Glenfeshie. Peter Fraser, the head of the Scottish Gamekeepers Association and a keeper on the neighbouring Invercauld Estate, said: 'Much of the problem is man-made and the result of stupid policies. There is abuse of deer because of unfenced forests and abuse of public money. Someone should be made accountable. They have had very, very poor results with regeneration at Mar Lodge.' The implication was clear: you can have your pinewoods, if you want. But only if they don't affect our deer.

In 2010, following reductions in deer density, Shaila's seedlings topped 20 centimetres. But in the face of such determined, vocal, powerful opposition, it wasn't enough. That year, the Trust signed a Section 7 agreement with SNH, a voluntary agreement to reduce deer numbers to a certain level. It is the precursor to a dreaded Section 8

---

‡ This is by no means a universal magic deer density, simply the one that worked at Mar Lodge. In other places, regeneration occurs at higher densities.

order, which gives the agency the power to come in and manage the deer themselves. Echoes of the controversial cull at Glenfeshie in 2004 may have focused minds in the Trust's upper echelons. In 2011, buffeted from all sides, the Trust decided to undertake an independent review of their management. This took evidence from staff, neighbouring estates, the local community and conservation experts about how Mar Lodge was being managed. It was thorough, and did not make for particularly comfortable reading.[14]

The report largely agreed with what Shaila and other workers had experienced on the ground. It found that 'the long history of heavy culls, with little regeneration to show for it, caused concern within members of staff as well as amongst neighbours and the local community'. It also found fault with decision-making and communication within the Trust, which 'led to a significant degree of mistrust'.

I ask Shaila about the review. It is clearly still a difficult thing to talk about. 'Communication's a really easy thing if you're doing something everyone's keen on. But it's much more difficult when the message you're trying to get across is one that people might not like. The Trust could have communicated better what we were doing, and been proud of what we were doing and [been] strong in our convictions.'

Importantly, the review found that '[sporting and conservation] are not irreconcilable in any way; rather it is a question of the timescale over which the objectives are to be achieved for the whole estate, and the ecological and land management context in which these are pursued'. The review made eight recommendations, which were all taken up by the humbled Trust, at least to a point. The review made two practical recommendations: that a four-kilometre 'strategic fence' should be erected, and that experiments in more intensive management techniques to jump-start woodland regeneration should be made. More, it said, could be made of sporting potential of the estate. It also made the point that the Trust's objectives were no less valid than those of their neighbours, and that if other estates were concerned about losing deer to the Mar Lodge vacuum then they could 'help themselves' through changing management practices on their own property.

'It really did change the relations and communications with our neighbours for the better,' Shaila says. 'Our neighbours wanted to have an opportunity to have their say to people who were important enough to make change. The sporting neighbours felt that they had an opportunity to have some input into what went on at Mar Lodge. And it also opened up a lot of channels of communication.'

Under a new head keeper and a new property manager, the team of stalkers, fiercely proud of their gamekeeping heritage, set about their task with renewed vigour. New strategies were implemented. The four-kilometre-long, off-set electric fence recommended by the review was installed. This would be low impact: it would stop deer moving into the regeneration zone from the south-west. In the woods it was off-set electric, two parallel, low fences which would be too wide for deer to jump over, wouldn't kill woodland grouse, and would allow the movement of other creatures. It was in effect a selective semi-permeable membrane. It also helped to ease relations with other estates. It showed that the Trust was capable of listening to criticism and willing to work collaboratively. As Robert Frost says, good fences make good neighbours.

In 2012, the little bar on Shaila's graph topped 25 centimetres. That year, following a mea culpa from the chairman of the Trust, who apologised for the Trust's handling of the 'deer question' and suggested that deer culling had been too heavy, a group of conservation charities ploughed into the fray. Adam Watson, always an outspoken voice to power, said that 'weak, unscientific attitudes have undermined the NTS staff for no good reason other than caving in to irrational and misguided political pressure.'[15] Conservationists were concerned that, just as the woodlands were beginning to recover, the rug was being pulled from under them, and that once again the landed elites would get their own way to the detriment of the nation. More than this, they could now point to the successes that were beginning to appear at Glenfeshie for evidence. It seemed that, in trying to please everyone, the Trust was on course to do the opposite.

The stalkers continued their task, refining and reworking as they went. They focused on deer hotspots, keeping up a continuous presence in the woods that kept the deer moving, creating a 'climate of fear' that stopped deer from picking at the best spots. By shooting at night and out of season, they could imitate the hunting techniques of wolves and, in the case of the smaller roe deer, lynx. It was tough work. Gradually, the clamour died down, as clamours tend to do. Mar Lodge and its staff tried to stay out of the limelight.

By 2015 Shaila's graphs were beginning to look respectable. In fact, the trees had nearly doubled in size since 2012. By 2016, twenty years after the Trust took on the estate, the tide had turned. A detailed survey that year found 835 hectares of regenerating woodland: the Caledonian pinewood was set to double in size. In 2017, the estate was designated a National Nature Reserve. By 2019, the average height of trees in Shaila's plots was tipping 80 centimetres, with the tallest pines topping 2.5 metres. Many of these had shot up since 2008 – new seedlings, unencumbered by grazing, growing faster than the original seedlings which had survived – just – decades of nibbling. Trees grew more on average in one year between 2018 and 2019 than in the six between 2002 and 2008. There were 2.5 times more seedlings in 2018 than in 2002. In 2007, 11.2 per cent of seedlings were above vegetation height. By 2019, that percentage was 67 per cent. This was in spite of the fact that the average vegetation height had increased from 18 centimetres in 2007 to 26 centimetres in 2019.[16] By 2019, monitored trees were producing their own pinecones. Some regeneration on the estate was now getting on for 5 metres tall and more.

By 2019, it wasn't just the pines which were growing. Birches, rowans, willows and juniper were increasing and growing vigorously. Orchids, lesser and common twayblade, small white orchid, frog orchid, were beginning to appear in places they hadn't been recorded before. Wildflowers which were rare a decade ago were becoming almost commonplace: alpine meadow-rue, globeflower, Scottish asphodel.

In 2018, I resurveyed that same 3.7 hectares which had contained four seedlings above vegetation height in 1998, and forty-four in 2008, and found that it contained 892 seedlings growing above the height of adjacent vegetation. A further 150 seedlings had grown enough to be considered trees in their own right, the largest of which reached 4 metres in height. There was 60 per cent less browsing to the ground vegetation layer. There was more heather, more blaeberry, more wildflower species. Lesser twayblade, one of my favourite orchids, had increased by 233 per cent. Wood ant nests had changed shape to cope with the increase in vegetation height, becoming on average 55 per cent taller over twenty years. Deadwood was accumulating, providing habitat for more insects, and nesting sites for more birds. Other things were remarkable by their lack of change. Canopy cover was pretty much the same over twenty years.[17] Woodlands change very slowly and very quickly.

All of these results are fairly predictable. This is exactly what the conservationists demanding a reduction in deer numbers twenty-five years before had said would happen. And yet to see it happen is a bit of a shock. This is ecological processes unfolding on a landscape scale. It is nothing less than the mechanics of the universe writ large across thousands of hectares of land. It is all rather beautiful.

Another lull in the tree monitoring, as we readjust guide lines made of blocks of wood and string (some of the best science uses the most basic equipment), and save documents to multiple places. I think aloud that it is hard to understand why the work to save the pinewoods caused quite the ruckus that it did. Shaila considers for a while. Finally she says, 'Part of it when I look back was the novelty of it. One important thing to remember is how much of a historic perspective you have on the site. If you'd come when I did, you'd have thought there was a lot of deer. But other people had seen it with twice as many deer as I had. For Deeside, this was the first time that deer reduction had really happened, and people didn't really know what to expect.'

The pine/deer conundrum is not just a conundrum of ecology. Rather, it is a clash of cultures. Red deer truly are totemic in the glens. For well over a century, they have been the nexus around which generations of lives have been arranged. They are truly loved by stalkers in both life and death. When the Scottish Gamekeepers Association talked of 'abuse' of the deer, it was talking about a wider perceived 'abuse' of an entire way of life.

It was sixty years ago that Malcolm Douglas's experiments first showed the extent of Mar Lodge's 'deer problem'. It has taken sixty long years to reach our current position of native woodlands returning to the Highlands not because trees grow slowly, but because powerful people simply did not want it to happen, and powerful people tend to get their way, at least for a while. For many in positions of power, the idea of reducing deer numbers to let trees grow was, and remains, anathema. When you write it like that it seems a little foolish, and so it is. Some sporting interests, offended by the accusation of centuries of land mismanagement, and perhaps grown comfortable in their perception of the Highlands as their own vast playground, accuse conservationists of deliberately misunderstanding the complexities of running a Highland estate, and hating deer. Conservation charities, frustrated by the slowness of progress and the intransigence of some landowners, have come to the conclusion that regeneration efforts are being deliberately obstructed by some in the sporting community who have no love of pine trees. However, the painful experience at Mar Lodge has helped to show that sporting estates and sporting culture can play a major part in the recovery and regeneration of the Caledonian pinewood on a landscape scale.

The image of the monarch of the glen on the shortbread tin is pervasive, hegemonic even, and it has a real impact on the lives of thousands of people living and working in the Highlands. The process of questioning that image stretched the relationship between Highland sporting culture and conservationists to the brink.

Back in 2000, in a book celebrating seventy years of Trust conservation, J. Laughton-Johnston wrote presciently of Mar Lodge that 'nobody knows, for all its good intentions, just how successful or not this management will be. The trick will be to balance patience and action and to untangle and identify the specific factors that are encouraging or hindering regeneration.'[18] Looking back through the documents, it is remarkable how far progress has actually exceeded expectations. In 2011, the review panel concluded that the target of 800 hectares of new woodland may be achieved slowly over the 200-year management vision. As it happened, the target was achieved five years after the review was published. Give nature an inch, and it will take a mile.

It's lunchtime, and a plague of midges has descended. We quickly eat insect-spattered sandwiches through head nets and get about reapplying Smidge. Is the woodland safe, I ask Shaila? Are you proud of your work?

'There's an awful lot of pleasure in seeing the woodlands bounce back. It's really quite exciting, having been here ten years and seeing very little progress, to see the rate of change now in the trees – you can see the potential being realised. It's been such a focus that now there's scope to turn our attention to other things that are good to do.' The most immediate problem is now what to do about her quadrats. 'The trees are getting enormous, almost too big to manage. In the next few years we'll have to work out whether we continue with the quadrats, or maybe reduce how often we do them, and make time for other work.'

For now, though, there is a slight breeze, and we are relieved from the attention of midges, so we get back to measuring the trees. There have been pines here for thousands of years, and now their future is guaranteed, at least for a couple of centuries. So too is the future of all of the species that depend on them.

'Scots pine, 78.2 cm, 12 cm, single-stemmed, 50–70 shoots, no browsing, code is zero. Scots pine, 40.4 cm, 11.8 cm, multi-stemmed, 45 shoots, no browsing, code is 3 . . .'

2

Green Shield-moss

A lily is more real to a naturalist than it is to an ordinary person. But it is still more real to a botanist. And yet another stage of reality is reached with that botanist who is a specialist in lilies. You can get nearer and nearer, so to speak, to reality; but you never get near enough.

Vladimir Nabokov, *Strong Opinions*, 1973

Spring comes late to the Cairngorms. Our swallows and house martins return three weeks later than the first arrivals to the south coast of England. Our wood anemones don't really do anything until April. Lapwings don't bother laying eggs until the end of April, and even then they usually suffer through a few snow showers before the things hatch. Nevertheless, there is a hint of spring in the mid-February air as three of us set out into the grounds of the Lodge. The air is a balmy 8°C, the river full to bursting with snowmelt. The first spring songsters, crossbills, tits, mistle thrush, ravens, are in fine voice.

There are three of us: Shaila, Petra Vergunst, local botanist and bryologist, and myself. We're not far from the main house, the opulent Mar Lodge itself; we're hidden away in the dark, ornamental woodlands that surround it. We follow a printout and a GPS to an extremely uninteresting-looking log. Shaila squares it up with a photo she took last year. We've a lot of logs to look at, so it's good to have a visual cue. Satisfied, we stalk in closer, as though the log is planning to run off given half a chance. There, in a rotting fold, is a small yellow plastic marker. Next to the label, growing out of the dull, wet, green-tinged wood, is a tiny, bright green protuberance sitting on a red stalk. It is a sporophyte – the fruiting body of a moss. I mark a '1' on a printout.

I look closely at the sporophyte. It has a certain aesthetic charm, I concede. I know it is a rare specimen. But, searching my feelings, I'm more excited by this find than I think I should be. After all, it's only a moss, right?

With only so much time and effort and money to go around, there is a limit to what conservationists can and cannot protect. Environmentalists tend to try to do conservation work which will provide the greatest return on investment for their efforts. For example, saving pinewoods also saves the creatures which rely on them, so they provide a good return on investment. However, there are any number of factors which feed into which creatures are saved; things like the rarity of the creatures in question and the undercurrent of politics playing out behind the scenes. Sometimes things get saved simply because someone rich happens to take an interest in them. Sometimes creatures get saved because people will pay large sums of money to hunt them. Sometimes it becomes fashionable to save a certain creature, like clownfish, because they turn up in a hit film franchise.

One distinctly unscientific rationale that lies behind which creatures get saved and which are left to their own devices is known as 'non-human charisma' – that is, how much humans are attracted to a particular organism, and why that might be. There are lots of reasons why humans find some creatures charismatic. Panda bears are cute and fluffy, eagles are big and impressive, tigers are beautiful and dangerous, dolphins are clever. The term 'non-human charisma' was worked on by the geographer Jamie Lorimer. He uses interesting, difficult sentences like this: 'In the more esoteric realms of natural history, organisms gain corporeal charisma by virtue of how they afford desired affects of culture and epistemology.'[1] Indeed.

This is by way of explaining how one of the big winners in the battle for non-human charisma, alongside the cute and rare Scottish wildcat, the weird and rare capercaillie, and the mighty Scots pine, is a moss which almost no one has heard of, *Buxbaumia viridis*,[*] the green shield-moss.

Mosses are bizarre things. They are plants, but they grow and multiply in very different ways to more 'complex' plants like grasses and wildflowers. They are ancient things; mosses first appeared in all probability over 500 million years ago. *Buxbaumia viridis* is even more bizarre than most mosses. It is a moss without any real leaves to speak of. Instead, it just sort of exists as a weird, inconspicuous greenish tinge. When conditions are right, the greenish tinge shoots up a sporophyte, which will eventually explode, sending spores out into the woods.

Like many mosses, *Buxbaumia* thrives in places which are wet and rich in nutrients. For this reason, like many mosses, it is particularly fond of dead trees, specifically ones which have fallen over. Find a dead, fallen-over tree, and you'll be looking at some prime moss real estate. The problem is, competition for deadwood is high, and at some point in its long evolutionary history *Buxbaumia* found that its neighbouring mosses, bruisers like *Hylocomium splendens* and *Pleurozium schreberii*, were a little boisterous, and prone to taking all the best places themselves. So, it did what species always do when there's lots of competition for resources: it specialised, and found itself an ecological niche.

Buxbaumia, over the course of the interminable ages that determine the course of evolution and speciation and whatnot, came to like wood that is soft and almost crumbly. It likes wood that is fairly mushy, but not so mushy that it gets too colonised by other mosses. It likes the wood that is pretty wet, but not too wet. It needs

---

[*] Bryologists, the people who study mosses, aren't really fans of vernacular names. In fact, they tend to make a point of avoiding them. So, for their sakes, we'll switch to the binomial for this chapter.

a constant supply of deadwood, because the stuff that it lives on is only suitable for a few years, before it completely decays or is taken over by other mosses and plants.

In the grand scheme of things, this isn't really a niche. In fact *Buxbaumia* isn't really all that specialised at all, because for almost the entire history of the great boreal woodlands of the Northern Hemisphere, fallen dead trees in dark, wet woods would have been absolutely everywhere. For this reason, *Buxbaumia* has a circumpolar distribution. A few thousand years ago, *Buxbaumia* made it to Scotland, following the woods as the ice retreated. For pretty much all of the time since then, *Buxbaumia* was, presumably, absolutely everywhere that there was woodland, particularly dense, wet, dark, pine-dominated woodlands. So you can imagine the surprise for *Buxbaumia* when suddenly, over hundreds of years, the deadwood resource declined.

The first problem was that pretty much all of the woodland in Britain got chopped down. The next problem, once we had grown some more, was that we decided we didn't like deadwood in our woodlands: it looked untidy, and it burned well as fuel, so we got rid of it all. By the time that people got round to noticing *Buxbaumia* in Scotland it was 1847, at which point naturalists were scouring the country, noticing obscure plants and species and dutifully recording them all in the proper Victorian fashion, and it was a decidedly rare moss. It was noticed a few more times up to 1887, and then no one noticed it again until 1951. In 1999, the total known British population of *Buxbaumia* was 31 sporophytes found on five logs in two locations in Central Scotland.[2]

Mosses can be fiendishly difficult to identify. There are over a thousand species of bryophyte (mosses and liverworts) in Scotland, and most will require a couple of microscopes and a hand lens and a complicated book to identify. It takes a special kind of person to get into mosses. Fortunately, there are a few of those special kind of people knocking about. In the early 2000s a small group of bryologists, notably Gordon Rothero and Andy Amphlett, attracted by the peculiarity and the rarity of *Buxbaumia*, put in a decent amount of effort to try and notice it a bit more. And they came up trumps. Quite a lot.

Stewart Taylor, one-time manager of RSPB's flagship nature reserve at Abernethy, now retired, has been doggedly pursuing the moss around the Highlands since 2007. He insists he's not a bryologist, 'just a dabbler', as many naturalists are, with interests in a wide range of species. 'It's an unusual moss, and it seemed like quite an interesting one,' he tells me, in the same reasonable tone that a car enthusiast might talk about an Alfa Romeo Spider.[†] 'I have a habit of looking for things that are very unusual. In the early 2000s, when we thought it was extremely rare, I was shown it by Gordon Rothero, on a log at Abernethy. Having seen it on this log, I spent a day looking for it in what I thought was similar-looking places. I think I shocked Gordon by finding about ten new locations that day. In that year I got about 140 locations for the moss.' By 2009 the small but growing band of bryologists had found 175 populations, simply because the right people had taken an interest in it.[3]

Meanwhile, behind the scenes, conservationists were moving to obtain legal protection for this new moss-de-jour. It was thanks to the fieldwork of David Genney (NatureScot bryologist extraordinaire) and Andy Amphlett, and others, that *Buxbaumia* became listed as a species of interest by the EU. This is a nerdy way of saying there was now more of a legal impetus to make sure that there were healthy populations of it. News got around the ecological grapevine that there was a moss which was rare, reasonably easy to identify, with European protection, and that it was under-recorded, so could turn up just about anywhere. This proved to be a catalyst for scientists. Says Stewart, 'If people have an eye for it and the habitat then they might find it. It's almost like a colour of green on the deadwood. I'd say the places I know to stop and look at are sections of log with

---

[†] I might be prejudiced, but I'd say that being enthusiastic about mosses is far more rational than being enthusiastic about cars.

the right colour of green created by the decaying process. It's only then that the sporophytes are found. A lot has turned up. People are getting their eye in.'[4]

By 2014, news had gotten out so much that a small group of bryologists, mycologists, ecologists and naturalists decided to try and find it at Mar Lodge. They did, with relative ease, and the two sites where they found it are among the most productive for the species in Scotland.[5]

This, in short, is the stream of events that led to me crawling about in a particularly dark patch of woodland, counting preposterous green protuberances on logs, and getting rather excited by the whole thing.

The plan is to return to specific logs where the moss had been found before, with the assistance of maps, fixed-point photographs and GPS navigation, and record as many preposterous green protuberances as we can find. Then we would look further afield for 'new' colonies of preposterous green protuberances.

One of the ironies of *Buxbaumia* is that, at Mar Lodge at least, it actually seems to prefer non-native trees and woods to the Caledonian pinewood. So far, it has only been found in the policies, the woodlands around the Lodge itself, planted up by the Victorians with exotic spruces and firs. These dark, dense, wet woods, packed full of dying and fallen trees, seem to be just about perfect for *Buxbaumia*. We quickly find the first logs. After a couple of minutes, we find some sporophytes and set about counting and recounting them. It's wet, earthy work, and soon my fieldsheets are a mess. Stewart was right; after a bit you do indeed 'get your eye in' and can spot a potential suitable location from a fair distance away. Looking up from a particularly verdant patch of protuberances, I notice a treecreeper a couple of metres away. I can barely hear its high-pitched call and make a note to get a hearing test done.

If ecologists are perceived as being nerdy, enthusiastic, a little lacking in the social-skills department, then bryologists are the ones that other ecologists think are nerds. Petra was an all-round ecologist before she 'got into' bryology. A year or so ago she decided to 'learn mosses' in the way that other people decide to learn to speak French. For approximately similar reasons, too: self-improvement, for fun, and with an eye on professional development. But it's fair to say that she's developed a deep affection for mosses, in part because they are so unloved and overlooked. People like me can become excited by the rare and unusual mosses, but Petra can see real beauty and value in even the commonest, least remarkable ones. Now, she's something of a moss ambassador. She's been setting up moss meetings across North-east Scotland. It turns out (who knew!) that there is a small but real appetite for bryology in the area. Once a month or so her group meets up, about ten people in total, to get to know mosses a little better.

As we move to a different *Buxbaumia* site we talk about under-recorded species. Part of what drives Petra is this knowledge gap. 'Mosses are everywhere, a really important part of the environment, but people just aren't looking for them.' This is true: NBN, the National Biodiversity Network, hosts an incredible collaborative database of records of the UK's wildlife. Of its unfathomable 222 million records, 3 million are of bryophytes. This sounds a lot, until you find out that there are 163 million records of birds. Thanks to its size and population density, the UK has perhaps the most studied biodiversity of any country in the world. But there are still huge gaps in our knowledge.

The talk does indeed get nerdy. Petra's good on her mosses, much better than myself, so I check with her the identification of a few common species, to make sure that I am remembering them correctly. Finding sporophytes takes patience and good eyes and practice, so naturally it gets competitive. We get impatient when we go without recording *Buxbaumia* for a while, and then are concerned that we have missed them. Petra tells me that she's actually now moving on to studying lichens, which are, if anything, an even more obscure group than mosses.

The morning goes well. We find twice as many sporophytes as the year before, and a few 'new' populations on logs where it has not been found before. After a quick lunch we head out to the 'prize specimen', the jewel in the crown of Mar Lodge's mosses: the largest single population of *Buxbaumia* in the country. It is a large, fallen spruce tree, lying untouched now for at least ten years and fairly rotten. *Buxbaumia* grows right along its full length. We work slowly and methodically along the trunk, marking our progress as we go. It is remarkably tricky work. Counting small green protuberances in a carpet of other small green things takes a fair amount of concentration.

Back in the office, we tally up the scores, and I'm surprised how much of an emotional investment I have in the moss. It's faring well. Overall we are up 60 per cent on the previous year and have recorded the second-highest amount since monitoring started in 2014. The 'largest population' is down by three, to 119 sporophytes, while sporophytes have returned to some logs where none has been recorded for a couple of years.

If monitoring answers some important questions, then it also raises more: how did the moss get here in the first place? How does it disperse? How long will it hang about?

'There are still lots of questions about it,' Stewart tells me. 'The spores don't go very far – only a few millimetres. But you get the right habitat and it does seem to be there. And we're starting to see it in other places too. I've found it on wood ant nests, quite regularly. Just the other week I actually found it on a mossy rock.'

Shaila, always a big-picture sort of person, is sanguine about the knowledge gap: 'As a scientist, I get really interested by these questions. But here at Mar Lodge we don't have the time or resources to answer them all. Yes, we could spend hundreds of hours studying the dispersal methods of *Buxbaumia*, but then we wouldn't be able to do our other work. And at the end of the day, would that knowledge change our overall management approach here? Probably not. At Mar Lodge we try to be as non-interventionist as possible. We want to get the land to the point where it can look after itself, and we don't really need to be involved all that much at all.' Still, this has not stopped annual monitoring of the moss. Nor has it stopped staff rolling in 'new' dead logs, to sit next to the current moss-laden logs, to provide a continuity of habitat for the future. As Shaila points out, 'You've got to accept that there are winners and losers in the game of landscape-scale habitat management. But we can always make exceptions for the real rarities.'

Caring for mosses, or rather, caring about mosses, is part of the whole practice of caring for the environment. In the rush to protect the big species, the smaller ones which make up the vast majority of the species of the earth are all too often forgotten about. Fortunately for the rest of us, the Stewarts and Petras of the world are there to see the value in the things to which the rest of us are blind. Learning to care about mosses, indeed learning even what they fundamentally are about, is an extreme thing to do. It requires a deep affection for the natural world, a great deal of concentration, and a lot of time and effort. And yet there is a growing number of people in Scotland and further afield that are doing just that. Following work in Scotland, bryologists across Europe, right the way to the Tatra Mountains in Poland, were mobilised to find this bryological curiosity. And the story is the same across the *Buxbaumia* range. It remains rare, and endangered, but thorough searching has shown ecologists that it is actually far more common than people thought.[6] It is just that people weren't looking for it. Funny as it may sound, *Buxbaumia* has turned into something of a poster child for mosses, even being described as a 'flagship species' of rare and endangered epixylic (deadwood-liking) mosses. That's right, folks. *Buxbaumia* has more non-human charisma than other mosses, through its oddness and rarity and (I guess) beauty. Through these processes, it has gotten humans to care about it enough that they will work to conserve its future.

The shy and retiring world of bryology is a wonderful world to stumble into – beyond its practical importance, bryology sits comfortably in a wider British ecosystem of eccentricities. Noting and recording and listing and tallying obscure mosses at which most of us would never even know to bat an eyelid has many similarities with trainspotting and fixing up motorbikes and stamp collecting. But it is also a pastime which has a real power to improve the world for everyone.

It's taken a while for me to understand quite why I've gotten so excited about *Buxbaumia*. This excitement has not really filtered down into an interest in other mosses, I'm sad to say. Nor, for me, is there any particular charm in the species itself. Instead, I think, I'm excited by what it represents. It is a tiny, overlooked portal to understanding the enormous ecological richness of our landscape. Understanding what makes *Buxbaumia* tick opens up a whole new way of appreciating our landscape. And discovering that it is far more common than we thought just makes me wonder, what else are we missing?

3

Roe Deer

'N uair a théid e 'na bhoile
Le clisge 'sa' choille,
'S e ruith feadh gach doire,
Air dheireadh cha bhì e;
Leis an eangaig bu chaoile
'S e b' aotruime sìnteag

When he's startled to motion
he's as swift as your vision
with speed and precision
he speeds through each forest
without seeming exertion
he's nearest, then furthest!

Donnchadh Bàn Mac an t-Saoir,
Duncan Ban MacIntyre, 'Moladh Beinn Dóbhrain',
translation Iain Crichton Smith, *Akros*, January 1969

Roe deer are the forgotten deer of the Highlands. They are not venerated like the stag. Lacking natural predators (beyond the occasional fox or eagle taking a fawn), roe deer populations are subject to the whims of humans. A couple of centuries ago roe deer were largely extinct in England, and on their way to extinction in Scotland, such was the lack of their woodland habitat and the rate of hunting. Now there are more roe deer in the country than at any time since the Ice Age. They are, incidentally, one of my favourite animals. Full of character, beautiful to look at, they lack the show-offyness of the stag. I've seen them high up the slopes of Carn a' Mhaim, but also in the parks of Aberdeen. I appreciate an animal that can thrive alongside humans. In this age of mass extinction, they bring me hope.

In 2018 I had a nice moment with a roebuck. I had finished working on one of Shaila's tree regeneration quadrats and had noticed a bit of fresh deer browsing on it. Walking back home, I came across some fresh roe dung. I scanned the area and slowed my pace and walked quietly. Looking up, a couple of minutes later, I saw the offending buck. He was a big lad, well fed, with nicely developing antlers and a rich coat. He was walking through a gorgeous patch of mature pines, snuffling around in bright green blaeberry. He completed the scene. I made a slight movement, he noticed me and shuffled off.

Using a gamekeeper's trick, I called him back by imitating a buck bark. Roebucks are pretty territorial, so pretending to be one is a good way of getting a response. Roebuck barks are a slightly alarming sound, particularly if you're used to the quietness of the average under-biodiverse British woodland. The buck stopped and turned. Curious, he circled round me, trying to sniff me. I kept barking, and he barked back, and continued to circle, forty yards away. Eventually, downwind, he realised that I wasn't a deer, and scarpered.

Back at the office, I told the gamekeepers where he was. The next day he was in the larder, ready to be sent to the butcher, and the pines in the quadrat were safe.

The difficulty of 'deer management' is that the conservationist is pulled between the desire to conserve wildlife in all its forms and the exigency of killing certain creatures so that others can flourish. It is an awkward, intractable position with no easy ethical answers, but which necessitates deep questioning from conservationists and sports folk alike. Mar Lodge staff suffered years of condemnation for killing too many deer, and indeed still do in some quarters. Others argue that killing even a single deer for whatever reason is barbaric. Still others have called for larger-scale culls at Mar Lodge. It is a curious paradox indeed that environmentalists call for the death of thousands of deer across Scotland, while the people killing thousands of deer across Scotland for sport are calling for restraint, and accusing environmentalists of 'abusing' the deer.

Such, then, is the controversy that surrounds the subject. But we are here on a quest to learn more about the land in which we live, and the experiences of the creatures with which we share this country, and to understand the people who work in it. To do this, we're going to need a mildly impressive piece of kit.

I meet Mar Lodge stalkers James and Dom on a cold, clear February night. There is a lot of snow on the ground. The mildly impressive piece of kit is a Hägglunds Bandvagn. Technically it's a tracked, articulated, all-terrain carrier. It was designed by the Swedish military to carry troops across the vast, snowy tundra, and so is perfect for carrying stalkers across the vast, snowy bogs of Mar Lodge. It's also fully amphibious, but I'm not sure that it has been used in that capacity here.

James and Dom hate the Hägglunds because it is loud and uncomfortable and has the unfortunate habit of breaking down in very remote places, in the dead of night, in a blizzard. This happened in 2018, resulting in a miles-long slog back to safety in the middle of the night with temperatures of −15°C. By the time James checked his rifle, it had frozen solid.

James and Dom also love the Hägglunds because it allows them to do their job more efficiently than their predecessors could have dreamed of. Even better, when it isn't broken down, it has a heater.

It took a long time to work out how best to protect the pinewoods. As we have seen, the original plan was to affect a slow reduction in deer numbers until a tipping point was met which allowed regeneration. Over the years, through Shaila's research and the experience of the gamekeepers on the ground, it became apparent that the worst time for damage to trees was in the winter and early spring, when the deer were forced off the high ground and into the woods for shelter. It was decided that licences for deer stalking at night and out of season were needed: legally, you can only stalk deer during the day and at certain times of year (the 'open season'), unless you have a licence to do otherwise. Nowadays, from December to March, the keepers are out a few nights a week, on patrol and watching for 'deer incursions'.

We set off across Doire Bhraghad, the Mar Forest, where the regeneration is at its thickest. James drives, Dom takes the lamp. I'm in the back with the dog, as far out of the way as possible. The lamp is a powerful torch which picks up the eyeshine of deer, sometimes several hundred metres away. Should eyeshine be picked up, James will jump out of the Hägglunds and take aim at the beast while Dom holds it still in the lamp. The Hägglunds is loud and bone-shatteringly uncomfortable. For such a large piece of kit, it is remarkably cramped inside. The tracks take in the snow with ease, but Dom warns about ice, on which the vehicle is useless, and snow drifts larger than a couple of metres deep, which can cause it to overturn.

Doire Bhraghad is quiet, as is Glen Lui, so we make our way into Glen Derry, slap bang in the middle of the regeneration zone. The lower reaches of the glen hold some of the finest woodlands on the estate. Now, for the first time in two hundred years, they are growing again, and it is only at the furthest reaches of the glen, far from large seed sources, that regeneration is still being stifled by browsing.

It is a clear night, but the lamp picks out frost forming in the air, peering through the serried ranks of thousands of young trees, mostly pine, but also a smattering of birch, willow and alder. The snow lies heavily on the seedlings, bending some of them almost in half. It is loose, powdery stuff, recently dumped by one of the big winter storms of the season. We reach the ford of the frozen Derry Burn, the Hägglunds plunges into the burn, reaches a surprising, bottom-tightening angle, breaks through the ice with a crack and a boom, and comes out the other side.

The lamp also picks out mountain hares. This is an endangered species itself; they've declined by an incredible 99 per cent in north-east Scotland in the last fifty years,[1] through a mixture of non-native tree planting and intensification of grouse moor management. The keepers keep a tally of all the hares they see: fifty in a night is not uncommon. Mountain hares are the most common prey item of the Mar Lodge eagles. Here, the hares are happy enough living in the woodland, just as they are in Scandinavia, but they are found right across the estate in varying numbers from the floor of the glens to the tops of the Munros.

We head past Derry Dam. Unless you know what you are looking for you wouldn't realise that there was a dam here. It was constructed in the eighteenth century to help remove the trees from Glen Derry. Logs would be held in the dam, and then floated down the river on a surge of water. It worked. Three centuries ago the woods stretched a couple of miles further than they do now. James spots a footprint in the snow. We get out to have a look. It was left by a red deer. It's a day old, he says, judging by the snow which has filled it. We're a few hundred metres beyond the woods now, further away from a source of seeds, and the regeneration is getting sparser here. Now that the woods are thick with young trees, this is the place which is most affected by browsing. The dog gets a quick run out. We stop for a cigarette break.

It's extremely quiet when the Hägglunds is turned off. The deep snow dampens all sounds, and there isn't a breath of wind. James talks about the winter of 2011 in which a Hägglunds wasn't available and in which scores of roe deer were pulled out of the woods. 'We were bringing back half a dozen a night. That was a pretty piece of work. We only had the Land Rovers back then, so we couldn't fill the backs with more than a few at a time. We had to drag them further as well. I don't think any of us knew there were so many roe deer in the place.' James is of Cumbrian sheep-farming stock, with a quick wit and generous to a fault. Like a lot of the Mar Lodge staff, he doesn't suffer fools gladly. He's been at Mar Lodge since he was a student keeper and came up for work experience. He is, by all accounts, something of a crack shot. It's no exaggeration to say that, alongside head keeper Chris Murphy, he's done as much as anyone on the ground to ensure the survival of the Mar Lodge pinewoods.

Dom is nineteen going on thirty-five and is also pretty handy with a rifle. He was a promising rugby league player at school until his career was stopped by a broken back and several bouts of concussion. He studied keepering because he loves the profession and has been shooting with his dad for most of his life. He has the bearing of a spaniel and, though I would never say it to his face, very occasionally he possesses a wisdom beyond his years. He also first came to Mar Lodge on a student placement. Now, he's a full-time stalker. We speculate, and we're fairly sure that when he was first employed here, aged eighteen, he was the youngest full-time stalker in the country. He loves deer. He is a one-man encyclopaedia on deer. He also loves killing deer. How does that work, I ask, loving them and loving killing them?

'There's the industry answer.'

And what's that?

'We'd rather we were doing it than people who don't love deer.'

What about his own answer?

'It's complicated.'

We discuss tactics. The Hägglunds is good. Very good, in fact. But it is not enough in itself to just drive around looking for deer. Some of the woods remain out of reach of even a tracked, articulated all-terrain people carrier. Others are too densely packed with trees to stalk through. We talk about high seats, which are used in heavily wooded areas in Europe, and remote cameras, squeakers and good old-fashioned deer stalking. Dom is critical of gamekeepers and stalkers who rely too heavily on their vehicles. 'It's just lazy, and the deer get used to it.' He sets high standards – he's hill-fit, able to walk up the steep mountain slopes as fast as others would walk along the high street, and enjoys putting students through their paces early on to assess their fitness. 'If they can't keep up, they can ruin a stalk.'

It's cold, so we keep going.

I watch on as they go about their work. They are polished.

They know each other well and work together well. They joke and whistle and sing songs. Sitting in the Hägglunds, surrounded by rifles and men in camo, I feel like I've accidentally enlisted in the army. Old-fashioned tactics come together with hi-tech kit. Copper ammunition is used instead of the traditional lead, as it is much better for the environment and indeed for the people eating the venison. James and Dom had been using lead ammunition until the Trust made the switch in 2018. They both admit that they were sceptical when copper bullets were introduced, but they have had no problems with them, beyond their much higher price tag than lead.

We stop again for a thermos of something hot. At a fairly deep level, lamping for deer just doesn't sit right with many gamekeepers and estate owners. It 'doesn't respect the deer'. It ignores the rules of the sport. It is cheating. Keepers who are happy to lamp for foxes, and even mountain hares in some places, balk at the idea of a noble stag or even a lowly roebuck meeting such an ignominious end. Ultimately, these deer are paying the price for the centuries of ecological destruction wrought by humans. It is difficult work and it lacks romance, but it's effective.

James and Dom are proud to be called gamekeepers. They both learned their trade at the same college. They take pride in their Mar Lodge tweeds (£1,500 a set), in the keepering tradition which they are continuing when they take guns out onto the moor to stalk stags and walk up grouse and catch salmon, when they return with guests to the gun room for a post-stalk dram of whisky by the fire. They pride themselves on toughness, hill-smarts, marksmanship, knowledge. The skills they employ in the Hägglunds were learned on the hill. But they measure their success in both regenerating pine trees and happy stalking clients. If lamping isn't 'real' keepering, then it sits with other activities which are. In fact, it goes both ways. The not-inconsiderable money that comes in from deer stalking with guests pays the wages of the keepers working to regenerate the woods. Nevertheless, James and Dom identify with a strong keepering culture, and express concern that it will be lost. Dom worries that, eventually, they will cease to be gamekeepers, and become simply 'rangers', 'like the Forestry Commission'. Keepering sits deep in the soul of a person, and those who would criticise the profession would do well to remember that.

We drop into the Quoich at midnight. It's a quiet night; they've been doing their job well. James and Dom discuss tactics again. Are the deer giving them the slip? An hour later, around the Elrigs* Burn, we spy a buck and a doe in the lamp. It's a tricky business: they're lamp-shy, and scarper as the light touches them. James can't get a clean shot, and the deer reach the safety of the thickest thicket. James and Dom are annoyed. They are paid to stalk deer, they say, so feel the pressure when they do not come back with a haul. I point out that it's pretty hard to kill deer when there are only a few about: gone are the days when 2,000 red deer and well upwards of 300 roe deer were sitting in the woods.

There remains a small population of resident roe deer making a living in the woods, grown 'too cute' for the lamp. They infuriate the keepers and impress them at the same time. Older, smarter bucks learn quickly to be careful responding to any old call, and their behaviour teaches other deer to be wary, too. Roes still nobble a fair few trees, slowing regeneration rates, alongside their larger red deer cousins. And the woods will always attract deer from other areas: there is much more to eat here than on the surrounding open moors. But now the growth rate of the regeneration is positive, and there are far more trees to nobble, so it's less important if a few trees get damaged.

As this style of woodland management becomes more and more common across Scotland, it is an ongoing concern for many keepers that they will 'shoot themselves out of a job' – that things will swing too far and deer will go from being overabundant to an endangered species. But the Mar Lodge experience shows that there will always need to be keepers on the ground, even with low densities of deer

---

* Elrigs in Gaelic is, suitably enough, a 'deer trap'.

in the woodlands. Besides, beyond the woods, there remains a herd of 1,500 red deer or thereabouts on the estate. The next problem, James says, is the regeneration itself: in some places it is already too high and thick to put the lamp through. That, we agree, is a very good problem to have.

In the long term, beyond the life-support system that it is currently on, the woods need the deer. There will come a point, sooner rather than later, when deer will be 'allowed' back into the woods. In time, traditional deer stalking with a twist will be offered to paying guests: roe and red deer stalking in the Caledonian pinewood. This will be a watershed moment – the true tipping point where Highland sport and healthy woodland ecosystems are aligned. There's talk of other projects, like community sporting lets, which would give locals who want to shoot for the pot the chance to do so while simultaneously contributing to conservation work.

An hour after the missed opportunity, we are back at the larder, cold and empty-handed. It is an odd feeling, and I'm not sure whether this is a good result or a bad one. I think about the buck and doe that shot off into the night, and how long they will get to enjoy the woodlands. I get to bed at 3 a.m. It is −10°C outside, and a light snow is thinking about falling. Dom and James will do the night patrols twice more this week.

On Thursday the Hägglunds sheds a track, and the two are forced to walk five miles in the snow. A week later, they bring in the buck and doe that we spied on our trip. Dom is pleased. He describes how they got the buck, imitates the 'thwuck' as the bullet ripped through the kill zone of heart and lungs, killing the deer instantly. They always process the carcasses the same night, bringing them back to the refrigerated larder, which is far warmer in February than the yard outside. The carcasses will go to the game dealer, and thence to the local butcher. The deer were gralloched where they dropped. The gralloch is the nasty bits that humans don't want. It is left on the hill for the eagles. Dom recalls a day a few weeks back when he surprised a white-tailed eagle off a gralloch, ten feet away.

Occasionally, about one time in ten, a beast in particularly poor health, beyond human consumption, will be left on the hill to return its nutrients to the ground, where it will provide welcome food for eagles, martens, the occasional hill fox. Such is the flush of nutrients that comes from a carcass that often small groves of trees will grow where the beast has fallen.[2] Each deer is weighed, numbered, sorted, and turned into the most ethically responsible meat that money can buy. It is wild, organic and more than carbon neutral, as its death allows trees to grow. It also happens to be delicious.

I think more about the contradiction of simultaneously loving deer and loving hunting deer. I wonder whether this is just a linguistic paradox; whether blunt words like 'love' are incapable of truly describing the complexity of the relationship between hunter and hunted. It is an older way of understanding and living with creatures than we are used to now, with our complicated taxonomy of pets for loving and livestock for eating and wildlife for watching and game for shooting and pests and vermin for exterminating. It is a contradiction that elevates the relationship between humans and wild animals beyond a superficial aesthetic appreciation of their wildness. It affords us a fuller knowledge of what it is to be alive and to live and die with other creatures. It respects the one infallible rule of ecology: we are all just something else's lunch.

But it is also true that the roe deer shot in the lamp are paying the price for centuries of unsustainable woodland management and for the extinction of apex predators at the hands of humans. 'Conservation shooting' remains at heart a utilitarian management tool. Roe deer die for the greater good of a balanced ecosystem, but it is a practice mediated by a complex system of imperfect humans. The act of hunting may aspire to an atavistic, pure reconnection with nature, a rewilding of the human spirit and a heightened respect for nature, but there will always be a cultural dimension to it too. Whether sportsperson, conservationist, or both, we forget this at our peril.

4
Woodland Grouse

At five in the morning, stumbling bleary-eyed through the woods, the philosophical movement that most resonates with me is existentialism. I think this is for two reasons. The first is that being awake at five in the morning, for whatever reason, is an inherently absurd state of being. The second is that, at five in the morning, one feels more keenly the physical sensations of being a human, even if that feeling only goes as far as the painful wrench that divides the conscious realm from the unconscious. There is the early morning ozone of clean air and sour, slept-on breath. I wonder whether there is a camaraderie among those who regularly wake up absurdly early, whether the bus drivers and bin collectors and early commuters give each other knowing winks, 'We're in this together, this is our time,' and scoff at us part-time early risers, who only occasionally emerge before the clock turns five to catch a cheap flight, embark on an ill-starred fitness regime or, in my case, look for birds.

It's April, and at five in the morning the sun is very firmly below the horizon. I'm tracking through the pines along a route I've walked enough times in the light to do without tripping over in the dark. It's cold, hovering around 0°C. I'm mostly listening. The best way to get the drop on the birds I'm looking for is sound. It travels well in the cold, still air. I hear a roebuck barking a few hundred metres away and ponder its future.

It's not until nearly six that I hear the sound that I'm listening for. It's a curious whattling, warbling, bubbling thing, interspersed with croaks and yells. It's a black grouse lek.

Crouching low in the heather, I creep forward. Grouse leks are notoriously easy to disturb. I crawl over to a small hummock and look down on the lek, a few hundred metres away at the edge of the woods. There are eleven cocks, and one of them is distinctly in charge. The males are metallic black and blue, with big showy tails, bright red eyebrows and the haughty demeanour of a duke whose monocle has fallen into the soup tureen. The females (known as greyhens) are smaller, chestnut, mottled, camouflaged. It's still early in the breeding season, and there don't seem to be any females. The frost melts into my coat as I wait a little longer.

Black grouse mating runs on a sort of 'king-of-the-hill' basis. Males will square up against each other, first with a series of choreographed dance moves and warbling croaking sounds. Then, if no winner is determined, they rip into each other. The winner gets to sit in the middle of the lek, which any onlooking females find extremely impressive and alluring.

At this lek, a big mature male, scarred and pitted about the face, is duking it out for the centre ground with a fresher, good-looking contender. Other side fights are taking place nearby. At the edge of the lek, there are two juvenile males, watching and learning. One plucks up the courage to have a punt at the middle. He is roundly seen off in a matter of seconds. Just as well there are no greyhens around. Their cooing call turbocharges male aggression.

Lekking sites can be used for decades. The actions of the lekking birds, all that trampling and jumping, keeps the vegetation beaten down and fit for purpose. Each bird is fairly loyal to his own lek site. That means that your average black grouse spends his entire life fighting the same males for the right to sit in the centre of the lek. Theirs must be a complicated relationship.

It is usual to describe the black grouse lek as a spectacle of nature, a glimpse into the lives of the non-human, a privilege to behold and a wonder of evolution. Perhaps it is the hour of the day, but I'm not feeling charitable. To me, today, the black grouse lek is the ultimate expression of absurdist existentialism. The whole scene seems preposterous, but nowhere near as preposterous as the fact that I've gotten up at five in the morning to see it and count its constituents, as indeed have half a dozen other members of staff dotted across the estate. It's a coordinated count, and we do them annually to see how the grouse are doing.

By six the sun is nearly up and the woods are alive with birdsong – tits, finches, crossbills, thrushes. A jay yelps, a woodpecker cheeps. I retreat to the office for thick, black, petrolly coffee and a plate of eggs.

Black grouse are currently doing fine at Mar Lodge. Their numbers peaked in 2013, following the spurt of vegetation growth that followed the reduction in deer numbers. Since then, following a series of wet springs, black grouse numbers have fallen back to roughly the same levels as the early 2000s. It may be that now the grouse could use a bit more grazing pressure to create the open clearings that they use for lekking and feeding. But broadly speaking, black grouse here are following the same pattern of population growth and decline as other populations across Deeside.[1] This 'fine-ness' is in contrast to the rest of the UK, where numbers have been in freefall for decades. Once a bird of farmland and the lowlands, like many species it is now confined to the uplands, and even here it is struggling in many places.[2]

Thing is, every time I'm out early in the morning, there's another sound I'm secretly listening out for. It is a different popping and wheezing and clopping, another ancient sound, altogether wilder and more mythical and complicated and bigger than the black grouse calls. It is the lekking song of the capercaillie, the world's biggest grouse species, which the Gaels called the horse of the woods. Worryingly, it's another bird that, until relatively recently, also seemed to be doing 'fine'.

Back in 2016, Shaila came into the office saying that there was a bird making sounds in the woods, and she was sure it wasn't a blackcock, and she should know, because she's been surveying black grouse and capercaillie for fifteen years. She knew it was a caper,* but wasn't quite ready to say it out loud. That's how rare they have become in Deeside.

I put it to the back of my mind. A couple of Saturdays later, not long after my morning coffee, along a gravel vehicle track, I stumbled on a monstrous bird stuffing grit into its gullet.† Turkey-sized, with dark, bottle-green and black plumage and a long tail, it flew into a tree and then, despite its size, melted into the woods. A ranger saw a pair of capers together a couple of hours later. Capers are particularly sensitive to being disturbed, so we set a few camera traps and left it at that. A couple of months later, an ecologist reported caper chicks in the area, and that was the last confirmed, successful breeding attempt of capercaillie at Mar Lodge. I've not seen a caper since.

Black grouse are birds of the woodland edge. They like trees, but they also like open ground. They used to be a common enough farmland bird. Back in the eighteenth century, Gilbert White wrote of black grouse in Hampshire.[3] By contrast, the allure of the capercaillie, like that of many charismatic creatures, is that it is rarely seen, a bird solely of the deep forest, a shadow in the woods. It is desirable for its rarity. Its predilection for being left alone by people acts as a marker of the wildness and otherness of the places where it survives. People are drawn to the Caledonian pinewoods because these special woods still harbour a suite of charismatic creatures, lost from much of the rest of Britain: capercaillie, wildcat, pine marten, red squirrel, crested tit. Visitors to the Caledonian pinewoods don't really expect to spot a capercaillie, but the fact that they *could* is enough to make the place special. But then, many of these creatures are common, abundant even, across the rest of their world range. If the caper's elusiveness hints at its uniqueness, then it is also a marker of just how much has been lost from our little island.

---

* Pronounced with a short 'a', as in 'capper' rather than 'caper'. Naturalists can be oddly snooty about how they pronounce species nicknames.
† Birds often 'grit', that is, eat stones. This is to help them break down food in their gullet.

Every time I head out early into the woods, I keep half an ear open for the popping, wheezing sound of a lekking caper. I'm yet to hear it. Every time I head out, I keep an eye out for their lozenge-shaped, oversized droppings, distinctively comprised of pine needles. I can't remember the last time I found any. Every year we receive about a dozen reports of capers from members of the public. We discount almost all of them as misidentified black grouse, which are superficially similar to look at, particularly if all you see is a big black blob disappearing into the dense woods. But every year there are one or two more credible reports. Depending on who you talk to, the number of capers at Mar Lodge sits somewhere between 'a handful' and 'none whatsoever'. Personally, I'm inclined to think that the number sits at the lower end of the range. Whatever the current status of capers at Mar Lodge, the woods here do hold a small but interesting part in the long history of caper conservation in Scotland.

For a bird that signifies the wildness of the largest woods, capercaillie are remarkably dependent on humans for their continued existence in Scotland. You see, capercaillies have actually become extinct in the UK before. You'll remember that by the eighteenth century the total area of Scotland covered in woodland was hovering around the 4 per cent mark. Only the great pinewoods remained, and we've already seen how complicated all that has turned out to be. It wasn't enough to support a species which needs large tracts of undisturbed forest.

So, in the early nineteenth century, step forward the fourth Earl of Fife, owner of Mar Lodge, hunting mad, and always on the lookout for exotic sporting quarry. By 1827, Mar Lodge had been turned over mostly to hunting, with a new wave of forestry operations on the way. Crofters had largely been cleared from the main glens. The Caledonian pinewood had recently been felled in Glen Lui, as had the adjacent pinewood of Badiness, and they were not to return for 200 years. The last pulse of large-scale natural regeneration had taken place fifty years previously.

The earl decided to add a little spice to his hunting diversions at his big fancy country pile. So between 1827 and 1829, he reintroduced capers to Mar Lodge, presumably with far less paperwork than is currently needed for species reintroductions.[4]

As it turns out, all that paperwork isn't really a bad thing. In fact, it's designed to stop people doing the stupid sort of thing that the Earl of Fife attempted. Of the first batch of birds brought over from Sweden only one bird survived the journey. Bored and lonely, the caper spent the next year wooing the local chickens, and entirely unsubstantiated legend has it that chicken–caper chicks were indeed sired. In 1829, he was joined by a pair of female capers. These actually did rear young, but the lineage of the Mar Lodge reintroduction project stopped there, and by the mid 1830s that was the end of that.

Fortunately, nine years later another scheme in Aberfeldy put a bit more effort into the endeavour. Fast-forward to the 1970s and there were 20,000 of the things knocking about in Scottish forests in a triangle between Glasgow, Aberdeen and Inverness, enjoying the new plantations of spruce and Scots pine that had been established in the first half of the twentieth century by the Forestry Commission. To put that in perspective, there were maybe twice as many capers in Scotland back then as there are black grouse now. Their abundance led broadcaster Tom Weir to write, 'We are fortunate, because the capercaillie is in serious decline on the continent, while in Scotland, there is still enough natural forest to support it.'[5]

This was, unfortunately, the zenith of Scotland's caper population. By the 1990s their numbers had halved. By the early 2000s they were down to 2,000 individuals. In 2016, a national survey estimated the population at around 1,000 birds, almost all of which were to be found in Speyside, with rump populations in Deeside, Perthshire and north of the Great Glen.[6] Capers were fairly common at Mar Lodge until the 1980s. The gamebag shows around forty capers were shot in the decade or so prior to the estate being purchased by the Trust.[7] Nature writer Jim Crumley came across a lek of four capers in Glen Derry without really trying.[8] Now, they are a cryptospecies, extremely rare at Mar Lodge, and on the verge of total extinction in Deeside.

Molly Doubleday is the advisory officer for the newly formed, multi-organisation effort to save the species from extinction. As such, she is one of only a tiny handful of people in the UK who is working exclusively on caper conservation. Molly is funded by a whole raft of charities and government bodies: NatureScot, FLS, CNPA, RSPB, SF. She spends her time fighting the root causes of capercaillie declines and working out which conservation strategies are working and which aren't. Given the parlous nature of funding for conservation in the UK, the project is relying for 60 per cent of its funding on the National Heritage Lottery Fund (HLF), and this runs out in 2023. In April and May she spends pretty much all of her time in the field, chasing down capers, camping out in camouflaged netting near lek sites and waking up at absurd times in the morning to count them and then remaining hidden in place until they have all gone so as to not disturb them.

What happened, I ask?

She tells me (in a broad West Country accent that takes me back to my formative years in Devon) that it is almost impossible to formulate just how little capercaillies have going for them. 'They don't like crossing open areas, and over time their plantation habitat has become more fragmented. Another factor is deer fences. These caused very high mortality, as capers crash into them as they fly low across the woods. This is still an issue, but it's been gotten on top of as people have become aware of it.' She pauses for breath: clearly there are a lot of problems. 'Then there's the wet springs. Rain in June is a real issue for chicks – their feathers aren't waterproof, so they can easily become chilled and die. With climate change, this problem is probably not going to go away. Then there's predators, foxes and pine martens, to name a few! These are an issue, but we generally think it's more of a problem for small populations surrounded by open land.' Predators tend to have more of an impact on small, unhealthy populations than they do on big, healthy ones, and in properly functioning ecosystems native predators never really cause problems for animal populations.

Then there's the impact of people drawn to enjoy the woods, in part by the capers themselves. 'It's just people accidentally disturbing them or going out to look for them. Very recent research suggests that capercaillie avoid woodland within 125 metres of busy footpaths and tracks. We don't know exactly the impact, but dropping surveys show there are fewer of them in woods with lots of people and their dogs.'

Capers, then, have very little going for them. But one way or another, it's all our fault. It's a sign of how reliant our woodlands have become on human intervention, and how all human intervention, even the stuff that is meant to do good for the forests (like fencing), causes all sorts of weird, unpredictable and often detrimental knock-on effects.

At Mar Lodge, people have been busy. Fences have been removed. Plantations have been restructured. Rangers tell people to keep their dogs under control through the breeding season. Now that there are fewer deer, the shrub layer (the heather and the blaeberry) has grown taller and more luxuriant, providing more vital insect food for chicks. But this brings its own problems with wet weather – chicks find it hard to dry out in long vegetation, and so chill and die.

Conservationists have been working for decades to halt the decline of capers. It may not seem like it, but their work has been effective: it has stopped capers going extinct in the short term. The Strathspey population seems to have stabilised. The work has also shown that we need larger areas of connected habitat to keep our capers into the future. Only the continued expansion of pine-dominated woodland will truly save our capers. But until that happens a new approach is needed to pull the species back from the edge of extinction in Scotland.

This is where Molly gets excited. The Cairngorms National Park is fronting a big new push with a big new idea: community-led caper conservation. The idea is to address a criticism that is regularly (and not always unfairly) levelled at conservationists; that environmental conservation is often too 'top-down', led by conservation organisations and government, with specialists parachuted in who lack local knowledge or sensitivity to local issues. So now the plan is to go heavy on that dreadful phrase which makes very cool stuff sound mind-numbingly dull: 'community engagement'. The idea is to push towards helping local people take ownership of local conservation issues and drive the systemic societal change from the bottom up. Most human disturbance of capercaillies is unintentional. Given a chance, people tend to want to look after wildlife rather than destroy it, just so long as it doesn't impact on their lives too much. The project aims to work closely with communities to educate people about the wildlife on their doorsteps, and to help people make the most of living with rare and exciting creatures.

Historically, the locations of caper lekking sites have been kept as jealously guarded secrets, so as to minimise disturbance. Molly says that 'because they're so vulnerable we've been secretive, but we've realised that it's not really working because human disturbance is increasing and it's unintentional.' But now, with more people than ever enjoying the benefits of our forests, and more people than ever wanting an encounter with a charismatic, rare creature, and with more people than ever owning dogs, being secretive doesn't cut it anymore.

In an age of environmental destruction, we all need to learn to live better with wildlife. But we also need to take as much enjoyment as we can from living with the wilderness. Thinking of nature as something 'out there', over which we have no impact, serves only to alienate ourselves from the landscape which is our home. Thinking instead of nature as an integral part of our lives, around which we can organise our day, and which we can play an active part in conserving, brings both greater understanding of and greater affection for the creatures with which we share our planet. The new conservation measures do just this, while also taking the best of what we have learned over decades of caper conservation.

It's early days, but big problems call for creative solutions. For now, these new solutions are being put in place alongside also some of the more 'traditional' conservation work. Molly's project is also funding habitat improvement works, sampling the genetic health of the caper population in the Cairngorms National Park, and improving monitoring methods. There's talk of reinforcing the Scottish population with more Scandinavian birds. The thought is that, given the small (far too small by current legal standards) number of birds released back in the nineteenth century, the genetic health of the Scottish population may not be great. By bringing in more birds, conservationists can both increase the long-term genetic viability of the Scottish population, and also bring in more, much needed, breeding stock.

What does this mean for Mar Lodge? Capercaillie are just about clinging on in the forests of Mar Lodge. In the long run, the habitat for them here is going to continue getting better and better, as the woodlands expand and join up. Conservationists have done all they can, bar a reintroduction. For them to thrive at Mar Lodge once more, it is up to all of us to leave them alone to do just that.

Three weeks after my first visit, I return to the black grouse lek for the second, and final, survey of the year. Another early morning, another existential crisis. Two blackcocks are still fighting it out for centre stage, as they have done here for centuries, while another three or four look on from the side lines. A bird flies to the lek, directly above my hiding place. He drops in at the edge of the action, assessing his competitors. He puffs out his chest, inhales, and begins his demonstrations and provocations in earnest. Black grouse calls fill the silent glen. The lek is a point of frenzied activity against a snow-capped mountain backdrop otherwise so still that it could be a photograph.

A forest without wildlife is like a football stadium without the fans: a nice piece of architecture, but ultimately meaningless. It is the creatures with which we share the world that give the world meaning, and if they become extinct, then Scotland will become a lonely place indeed.

5
Kentish Glory

It is one of the last days of the field season, in late October, and Gabby Flinn, who is usually based in Speyside, is over on our side of the hill for a change. Gabby is an entomologist (insect scientist), but first and foremost she's a myrmecologist (ant scientist). Gabby loves ants.

It's a glorious afternoon, but it snowed in the night, and this morning snow was lying outside the office window. Hardly ant weather, but Gabby is undeterred. She has found our quarry at 2°C before.

The woods are looking gorgeous. The birches are gold, the rowans burnt red, the pines resplendent. The air is clear and crisp. We veer from the path into the woods, where the heather is deep and the regeneration is bordering on obnoxiously thick. It's a pain to walk through. A tit flock roves around the pines, and we marvel at the miraculous constructions of the wood ants: giant thatched nests of pine needles, grass and heather, laced through with vents and tunnels.

Gabby is the project officer for the Rare Invertebrates in the Cairngorms project, a collaboration between the RSPB and the Cairngorms National Park Authority. Her job is to save six species of particularly rare insect, and today we're looking for *Formicoxenus nitidulus*, the shining guest ant.

The project is looking at some very weird and wonderful creatures, but for sheer absurdity, the shining guest ant takes the biscuit. 'It's a parasite, we think,' Gabby explains, as we follow a grid reference to a wood ant nest. 'It lives in wood ant nests, but it's not a wood ant. In fact, lots of species do this – maybe a hundred or so insects that we know of do it in Britain. But the guest ant . . . it really is shiny, completely shiny all over. And tiny. Smaller than you think it will be. Shining guest ant has never been recorded in Deeside; if it were here it would be a major coup.'

I ask questions: what does it do in the nests? What does it eat? What kind of nests does it like? How many are there? Is it just really under-recorded?

The answer is always that same: 'We don't know. We know literally next to nothing about this species. That's why we're doing this work.'

Gabby is particularly keen to see whether we can find it in the nests of *Formica exsecta*, the narrow-headed ant. This is itself an extremely rare insect in Britain, known from only five locations in Scotland and a single site in Devon. Fifteen years of monitoring at Mar Lodge have yielded a maximum of twenty *exsecta* nests at any given time. Some of these are over twenty years old. Until now, the shining guest ant has only been found in the nests of *Formica aquilonia*, the Scottish wood ant. If it could be proven or disproven that the shining guest ant lives in the nests of species other than the Scottish wood ant then that would be a real point of interest (if you're into ants, that is).

The only way to find shining guest ants is to sit and look at the nests and wait for a small shiny blob to wander across your line of sight. Today the ants are clearly not pleased about the temperature. A writhing caterpillar is hauled into the nest by a group of workers in what looks like a video tape slowed down to quarter speed. *Formica exsecta* doesn't like cold weather – they are not as good at thermoregulation as other wood ants. We try a few nests, with no luck, so move on.

Most of the nests here are the monstrous constructions of *Formica aquilonia*. These are huge, intricate things, usually set up at the southern edge of a pine tree's canopy. They can easily be a metre in diameter and a metre high. Here, the ants are working at half speed, rather than quarter speed. These nests can last for decades, hosting tens of thousands of ants, whose total combined neurons are equivalent to the amount found in a human brain. Watching their comings and goings, shining guest ant or not, can be mesmerising.

Ants are one of the most integral parts of a functioning ecosystem. They constitute around 20 per cent of the earth's terrestrial animal biomass.[1] In the pinewoods, they control outbreaks of tree-damaging species. They are food for birds and mammals. They farm aphids to secrete honeydew. Go to a damp part of the woods, and you'll find that some of the pine seedlings have 'sleeves' of moss. These are created by ants as 'barns' for their aphid livestock.[2]

We talk shop. The project has had a good year. They have found new sites for all of the species Gabby has been working with and built up a huge network of volunteers with a newfound interest in insects. A big part of the project is about getting people thinking beyond the big fluffy mammals and rare birds, and towards the other important parts of an ecosystem. The vast majority of us don't give much thought to insects, which is a shame, because they are extremely important. The Rare Invertebrates in the Cairngorms project is in part an attempt to redress the balance – to get people excited about insects.

I mention I'm writing a book. 'You have to write about shining guest ant – it's the frontier of ecology – we know *nothing* about it. And don't forget about *Blera fallax*, that's a good story. And I'm sure there are some of them here somewhere.'

It's a reasonable story. *Blera fallax*, the pine hoverfly, was first discovered in Britain in Braemar in a hotel room in the Fife Arms. A visiting entomologist heard a buzzing and swatted it away, only to discover that it was a hoverfly species that had never been found before in the country. The pine hoverfly is probably the rarest of Gabby's target species, so she and hoverfly expert Steven Falk came to look for it at Mar Lodge in 2018. They didn't find it, but they did find five other species of insect which were previously unrecorded on the estate. So perilous is the status of pine hoverfly in Britain that Gabby's team have taken larvae from the wild to hand-rear them for future reintroductions. The larvae are now being grown on by folk at the Highland Wildlife Park in hummus pots.[3]

There are many cool, rare insects at Mar Lodge, each of which deserves a chapter in its own right. Like the species of small empid fly discovered at Mar Lodge that has never been recorded anywhere else in the world.[4] Then there's the pearl-bordered fritillary, a truly stunning, rare-ish butterfly that is one of the few butterfly species bucking the trend of national decline, thanks in no small part to the work of conservationists. There's the mountain mason bee, found at just two sites in Britain. There's the five-spot ladybird, a resident of inland shingle which likes braided river systems, and the silver stiletto fly, which has lekking sites on the same habitat. There's the mountain burnet moth, found only on hills of a very specific size and with a very specific mixture of crowberry and boulder scree, and known from around six spots north of Braemar. There's *Callicera rufa*, another hoverfly and the star of Fredrik Sjöberg's beautiful, esoteric meditation on all things hoverfly, *The Art of Flight*.[5] So this chapter is by way of apology to the many entomologists out there, who are very used to getting short shrift in this sort of book. But I've decided to write about moths, and in particular about one which has never actually been recorded at Mar Lodge, the Kentish glory.

If you want to understand the parochial yet grandiose nature of, well, naturalists, then look no further than the naming rights of British lepidopterists, that is, butterfly and moth people.

Here is as rich a vein of poorly named creatures as you will find anywhere in the world. The Camberwell beauty, for example, was once recorded in Camberwell, and never again. It is, by contrast, extremely common in southern Europe. It's the same story for the Bath white, whose other name, Vernon's half mourner, seems far superior to me, no matter who Vernon may have been, or why someone was only half bothered to mourn him. The Scotch argus, admittedly, makes a fair amount of sense, but is also found across Europe. Likewise the Lulworth skipper, which in Britain is indeed confined to a tiny area around Lulworth, but is one of the most common European butterfly species. Such nomenclaturial perversion reaches its apotheosis in the Arran brown, which has in all probability never actually been recorded on Arran, or indeed anywhere else in Britain for that matter.* One is reminded of the infamous headline: 'Fog in Channel: Continent Cut Off'.

The Kentish glory suffers from a similar affliction. This large brown moth, whose only current known British populations are in North-east Scotland, was admittedly once found in Kent, and presumably the length and breadth of the UK, but it hasn't been seen in England, let alone Kent, for over fifty years.

The problem is that the Kentish glory is particularly fond of young birch trees. It is a creature of scrub, a much-maligned, much-tidied-up habitat of low bushes and small trees. Scrub is a successional habitat. It is the bit between a field of grass or an open moor and mature woodland. Being a creature with a fondness for scrub was fine back in the days when birch trees were sprouting up, growing and dying across the whole country, and the to-ings and fro-ings of grazing species and their predators (man and beast alike) formed an infinitely complex mosaic of open spaces, closed canopied woodlands and all points in between. But these days that simply doesn't happen. Now, we tend to define a habitat as what it was when we first noticed it and either do our utmost to keep it that way or else turn it into a car park. And birch trees don't stay young for very long, and Kentish glory moths don't travel very far. Like a significant proportion of our endangered species, Kentish glory moths like large-scale, functioning, complicated ecosystems, and these ecosystems don't really exist anymore in Britain.

Now I must admit to being slightly nonplussed by moths. It's

---

* This was, apparently, due to the entomological equivalent of an accounting error.

not that I don't like them, far from it. It's just that there are lots of them, too many for me to get my head around without then forgetting about other important things. There are about sixty species of butterfly in Britain, which is a very manageable amount. There are, by contrast, around 900 species of 'larger' moths, and a frankly obnoxious 1,600 species of micro-moth, which are all very small and very brown and very hard to tell apart. They also, and this might be stating the obvious, tend to hang about at night, which is when I tend to be asleep.

I have seen a Kentish glory, once, in Poland, on a birch tree, at three in the morning. It was quite nice. It was quite large, as far as moths go. It takes its limited palette of brown, darker brown, red brown, black and white about as far as it will go: nature as designed by an abstract expressionist. I suppose it is as glorious as a medium-to-large-sized brown moth can be.

Patrick Cook, ecologist at Butterfly Conservation, is by contrast very good at moths. By April 2019 the national hunt for Kentish glory, spurred on by Gabby's project, has reached Upper Deeside, and he's made the long trip up the road to see whether we have any here. It's a long punt – the last record of Kentish glory in the area comes from 1980, a couple of miles from Mar Lodge land, but there is now a considerable amount of good habitat for the moth here. While I'm not massively 'into' moths, I have enough of the collector's mindset in me to enjoy seeing rare things, particularly if they form part of an intricate piece of machinery like a functioning ecosystem. I also have a sneaking suspicion that Mar Lodge is the greatest place in the country and like to see this proved by its having lots of rare creatures about. So, I have skin in the game.

Patrick's in the right job, because he clearly loves moths. He's somehow managed to retain both the vernacular and binomial names for pretty much all the species, and he's also somehow managed not to forget about other things like plants and birds and such. Like many conservationists, he has put his time in across the country in volunteering projects and short-term contracts, working in Shetland on red-necked phalaropes (which are birds, not moths) and other such exciting creatures. Now, like many conservationists, he works extremely hard in the name of saving the environment. He does this because he feels duty bound to do so, because there is no other choice in such an underfunded industry where there are scores of qualified people for every paid position, and mostly because it is his passion. From April to August, peak time for moths in Scotland, he racks up hundreds of hours of overnight moth-trapping surveys. We've done well to get a whole day from him at this time of year. 'I love this time of year,' he says, somewhat bleary-eyed. 'It's when I feel most alive.'

Fortunately for me, the Kentish glory is a 24-hour-a-day of moth, so we start our day at a sedate time in mid-morning with a tour of places I've earmarked as potentially suitable sites for the moth. Pleasingly, Patrick's not entirely in disagreement about my site selection. The reason we think there's loads of suitable habitat around is that the Kentish glory likes very young birch trees, and we've got hundreds of thousands of those. We stop at a particularly promising patch, where the birches are just the right size. Patrick pulls out three tiny plastic lozenges, ties them to trees spread a few metres apart, and opens their stoppers.

'These are lures. They're basically comprised of female Kentish glory pheromones. It's breeding season for KGs,' (like birdwatchers, lepidopterists also have all sorts of street words. Kentish glory is usually referred to as 'KG') 'so the males are attracted to the female pheromone. If there's a male within a couple hundred metres of here, we'll smoke him out.'

'Oh, right. And where does the pheromone come from?'

'They make it in a lab.'

'They what?'

'They make it in a lab.'

I must look a little lost, so Patrick unpacks it a bit for me. 'First they took a female KG in heat and stuck her in a little tub for a while, then they released her and sucked out all the gas from the chamber and isolated the pheromone. Then they stuck that in a mass spectrometer (a sci-fi device that tells you the chemical composition of things by the light that they reflect) to work out its chemical properties. Then they synthesised it in the lab.

'Of course, it wasn't as simple as all that. My colleague Tom Prescott spent two years working with Canterbury University working out how to actually use the pheromones to find males. We tried a few different ways. We even included female moth puppets to see whether that helped,[†] but we're pleased with the system we've got now.'

We wait for ten minutes for a randy moth to fly in. No luck. We do this several times across several square kilometres. We hear hundreds of willow warblers, more than I've ever heard before, and scores of tree pipits and redstarts and tits and siskins. Patrick tells me of the bumper hauls you get when you stick a moth trap under a sallow. He tells me how they really are a hugely important, underappreciated group of pollinators, and that they are a big part of what all these willow warblers are eating. But we don't find any KGs.

All is not lost. Gabby and her volunteers find KGs at ten new sites within the National Park in 2019 alone.[6] Another arm of the project is looking at captive rearing, to release the species into suitable new areas. Meanwhile, Patrick has given me a list of other super rare moths that he thinks are probably lurking about the place somewhere. My homework is to find a few of them.[‡]

The woodland restoration efforts at Mar Lodge seem to have come too late to save the Kentish glory moths which would have once undoubtedly graced the estate's birchwoods. It has nearly come too late for another species which we have already met – the narrow-headed ant. This is an ant of woodland edges; like the Kentish glory, it seems to like the bits that are somewhere in between open moorland and closed-canopy woodland. By 1995, the population had been reduced to two areas at Mar Lodge, which were too far apart to

---

[†] It didn't.
[‡] I failed at this homework, miserably. Fortunately, Patrick came back to look for a couple himself and found one species quite easily.

interact with each other. For two decades we have watched the population wobble along, losing nests here, gaining nests there. But it is reduced to such a low ebb that it doesn't seem to be able to disperse to make use of this new habitat. To make matters worse, this is a species that benefits from a bit of grazing – the conservation work which has saved the pinewoods in the long term may be damaging the fortunes of one of its species in the short term.[7] Jenni Stockan, entomologist extraordinaire from the James Hutton Institute, has been following their fortunes for the last decade. In 2019, she undertook genetic analysis of the Mar Lodge narrow-headed ants to see how healthy they were.[8] This work is the precursor to a potential translocation project: moving ants to new, suitable locations, and reinvigorating the genetic viability of the population by introducing narrow-headed ants from further afield. A similar project is already underway at the only known site for the species in England, funded, as these things usually are, by the Heritage Lottery Fund.[9]

Synthetic pheromones, genetic analysis, detailed GIS software, hummus pots full of hoverfly maggots and an army of volunteers are the new face of insect conservation in Britain. These sorts of elegant but round-the-houses approaches to saving creatures, a mixture of hugely clever and slightly improvised, are common in the conservation world. There is something endearing about the small band of nerds, this happy few, that devote their weekends to saving tiny brown critters that most people simply do not notice. All incredible stuff, undoubtedly.

Around 2,000 species of insect have been recorded at Mar Lodge. The number grows every year, as we get better at recording them, or specific projects like Rare Invertebrates in the Cairngorms provide the funding and interest for another round of species finding. But it makes you wonder: the lengths people have to go to save a moth, or an ant, or a hoverfly; how bad have things gotten?

One of my daily tasks is to man the moth trap. It consists of a powerful light on a clockwork timer and a long drop into a poison-filled receptacle (more death). Most days, I empty the automated trap and dispense the contents into little packets. I then send these slightly suspicious-looking packages off to staff at Rothamsted Research, who identify all of the moths, midges, wasps and other delights ensconced inside. It's fairly grisly, unpleasant, smelly work, and a source of a certain amount of unresolved guilt on my part, even if there is literally no chance that the traps will cause species extinctions, or even have any sort of local effect on moth populations.

The great value of this trap-monitoring programme is that it has been running for nearly fifty years. It is also one of a network of eighty similar such traps dotted about Britain. It therefore provides a 'constant effort' survey of enormous proportions. It is, in fact, part of the largest insect survey in the world.

The results of Rothamsted's work are fairly terrifying. Over a 47-year period involving 7,593,437 individual records, they found that overall moth abundance declined by 31 per cent.[10] The data was also used to inform a study from 2006, which found that two thirds of Britain's larger moth species declined in abundance between 1968 and 2002.[11] An update of this study, published in 2013, found that 37 per cent of those species decreased by at least 50 per cent.[12] But there are also glimmers of hope. This study also found that one third of species became more abundant with fifty-three species (16 per cent of the total) more than doubling their population levels over that period.

It is no secret that insects are doing badly. The reasons are the usual culprits: pesticides, intensive farming, increased urbanisation, less habitat, less connected habitat, climate change. Above all, the Kentish glory shows us that insects need connectivity. Like the capercaillie, they cannot survive in islands of habitat surrounded by hostile territory, particularly when the islands of habitat are shrinking. Scotland's woods need to be bigger and more connected. But they also need to be wilder, and to behave more like dynamic ecosystems should. Humans can replicate these patterns through conservation management, but only imperfectly. The plight of the

Kentish glory and the narrow-headed ant and the redstart and the tree pipit and the black grouse show that we need to think of conservation not in a new way, but in a very old way – we need to give nature space to breathe and to do its own thing. It shows us that to have a truly functioning environment we need to think beyond the idea that nature lives in nature reserves and has no place anywhere else. It shows us that saving our most endangered species will take a whole new way of thinking about our landscape.

We can turn things around. But we need to do it soon. As Butterfly Conservation noted in a recent publication about the state of our moths, 'a focus on threatened moth species, while essential to prevent further loss of biodiversity, is not enough. Pervasive environmental degradation and the decline of common species demand the recreation of a rural and urban landscape that is much more hospitable to biodiversity. Carefully targeted and properly resourced agri-environment and woodland management schemes would be a significant step towards repairing Britain's natural heritage and safeguarding the ecosystem services that underpin human welfare.'[13]

We know what we have to do. Now we just have to do it. And in the uplands, it's the easiest thing in the world – we just need to let go and let nature take its course.

I have been wondering about the concept of the 'wild' and its defiance of easy definition. I wonder whether this defiance is actually a clue to its true meaning, whether it is best described by means of power and autonomy. Is being wild simply having the ability to do the unexpected or even the harmful, beyond the control of human interference? Is wildness equal to freedom?

Phenology is the study of species interactions with annual cycles. It's the arrival of the house martins, the harriers, the emergence of the first wood anemone, the first roar of a stag before the rut, the first flight of a pearl-bordered fritillary. There is another phenology, one far less poetical and far more pertinent to one's daily life. This is the phenology of irritants. Wasps make a habit of building nests in unfortunate places in spring – one nest in particular cured me of my fear of them. The nest was in my attic, and wasps kept finding their way into my living room and climbing up my trouser legs. After this had happened six or seven times, I realised that wasp stings aren't as bad as all that, and then a frost came along, and that was the end of the nest. Then come the ticks – no malign deity could ever contrive to design a creature more loathsome to me than a tick. Then there are horseflies, the small, grey clegs and the larger green-eyed monsters that arrive in warm weather in mid June, biting through clothing, leaving angry welts. After them come the keds: flatflies which cling and crawl and hide about one's person, revealing themselves only by a small, deeply unsettling scuttle over the back of one's neck, sometimes days after you have been anywhere a ked might be reasonably expected to be found. In July flies horde about in their hundreds, bringing something akin to claustrophobia to their targets. And, of course, then come the midges, the clouds of which, in August, can drive even the most stalwart of naturalists to distraction, and have, according to local legend, driven some unfortunates to madness.

The wild, then, is a place where car windscreens are still spattered with creatures at the end of a drive, where wasps still make a nuisance of themselves, where a day in the field will not be concluded without shaking unwanted visitors from one's hair and checking and rechecking various orifices for parasites. The wild is a place where life still makes your misery its business, and life is all the richer for it.

The woodlands are returning to their former glory at Mar Lodge, increasing in size and diversity and connectivity and complexity, providing homes for more creatures from ants to eagles. But this is a long-term project. The vision of a fully functioning woodland ecosystem, which dances to its own tune, and has the freedom to bite and sting, which supports humans rather than having humans supporting it, remains decades away. In the meantime, the fate of creatures like the Kentish glory are in the hands of the tiny band of people who are interested in the fate of the small, boring brown things that inhabit the earth, and without whom people could not survive.

6

On Patrol

> Are not these woods
> More free from peril than the envious court?
> Here feel we not the penalty of Adam,
> The season's difference, as the icy fang
> And churlish chiding of the winter's wind,
> When it bites and blows upon my body
> Even till I shrink with cold, I smile, and say
> 'This is no flattery. These are counsellors
> That fleetingly persuade me what I am.'
>
> William Shakespeare, *As You Like It*, 1.2

As Shakespeare's characters move from the 'envious court', the city of plotting and scheming and general nonsense, into the wilderness of Arden, there is an acceptance that the normal rules of civilisation and society have been upended. Arden becomes a place of 'true' reality, in which Shakespeare's characters test their true worth against an indifferent Nature. However, this inversion of reality is only a temporary one, and by the fifth act the characters must return to the city and 'normal life'. This is the standard trope of pastoral literature: getting back to nature may 'fleetingly persuade me what I am', but you still have to go back to your day job at the end of it.

The forest, then, has a long cultural history as a refuge from 'normal life'. Preas nam Meirleach, Robbers' Copse, in Glen Luibeg, is a well-named example of this. Cattle-rustling was a way of life in Upper Deeside certainly until the late seventeenth century. The copse is the last substantial piece of woodland in which prospective thieves could hide until you reach Rothiemurchus on the other side of the massif.

Back in 1996, just after the Trust had taken on management of Mar Lodge and was working to enhance its wild land quality, the geographer William Cronon was writing his influential essay 'The Trouble with Wilderness'. He argued, 'Far from being the one place on earth that stands apart from humanity, [wilderness] is quite profoundly a human creation – indeed, the creation of very particular human cultures at very particular moments in human history ... As we gaze into the mirror it holds up for us, we too easily imagine that what we behold is Nature when in fact we see the reflection of our own unexamined longings and desires.'[1]

This is the paradox that lies at the heart of that confusing, ancient trope of 'getting back to nature'. Nature has become a place to return to, rather than to live with – a holiday destination rather than a home. As such, the wild lands of Scotland, places like Mar Lodge, provide a condoned space for the temporary circumvention of everyday banality. But in so doing, do they not also reinforce our concept of 'nature' as removed from the everyday? More concerning still, if 'getting back to nature' is a philosophical quagmire, then it also has a physical impact on the landscape itself. After all, humans have a remarkable capacity for damaging and even destroying the things that they love.

It is comforting that our relations with the natural world were being questioned by Shakespeare 400 years ago. We've already seen through the work of conservationists that the 'real world' of nature is far more closely bound to the 'false world' of humanity than many would like to think, even in the remotest corners of the Highlands. The Mar Lodge pinewoods are no pristine wilderness: they have been a home for humans since the ice retreated. They've been forested and grazed by livestock, crofted, cleared, hunted through and protected, and now they are a retreat from the travails of modern life for tens of thousands of visitors every year.

This poses a number of difficult questions: how do you facilitate the means for people to see and enjoy and benefit from nature without harming it in the process? How do you cater for the multiplicity of different 'natures' that each of us holds in our respective and collective heads? Indeed, can people really get back to nature at all?

Kim loves talking to people about nature, which is great, because that is essentially her job. Kim is one of the rangers at Mar Lodge. She's been here for eight years and knows the place inside out. It's a Friday evening in early May. We load up the pickup with butts of water, litter pickers, shovels. Then there's binoculars, a handheld GPS, a Spot locator,* a radio, a camera.

'Now that the woodland is regenerating, the greatest danger to the woods is us,' says Kim. 'And a big part of that is the risk of wildfires. Most native woodlands in Britain burn like wet asbestos. But the pinewoods are the exception. For lots of people, the whole point of coming out to places like Mar Lodge is to do the wilderness thing. People expect to be making a campfire. Fire is fun. How often do people get the chance to play with it?'

We're heading out on patrol, as the rangers do here most days during the summer. Not far from the Lodge we pass a somewhat unsettling reminder of the long human association with the Mar woods. Kim points out a small, very dead, decaying mass of tree, tethered to the ground by stiff wires. 'This is the Gallows Tree o' Mar', says Kim. 'There's a good story about the last man to be hanged there. His mother cursed the Farquharsons, the landowners who had him hanged. By the turn of the nineteenth century, the Farquharson's male line died out.' Heritage is important – the tree has been kept in place by wires since at least 1925. We keep driving towards the Linn of Dee, and there's smoke wafting over the river. Kim sighs. We park up a respectful distance away and head over to the group,

---

* A fancy tracking device that allows other people to know where you are when you are working alone in places with limited phone reception.

young professionals, Instagrammy types, a couple of hipster beards. It's an idyllic scene. They're getting back to nature, and we're going to ruin their fun.

Kim is good. Her Cumbernauld accent is simultaneously friendly and firm. She tells them there's a high fire risk, and that they shouldn't have a fire. They apologise and help as she digs out the fire. She makes them put their hand in the hole to see how hot the peat has gotten. She explains that it's not just the trees which can burn, but the underlying peat. The soil is highly flammable when it is dry. Fires can sit under ground for days, before reaching back up to the surface. She says the ring of stones they've put around the fire is pretty, but it is also pretty much pointless. She dowses the fire site with fifty litres of water. The ground fizzes and boils. She's too polite to mention the fact that they have walked right past a 'No Fires' sign and completely ignored it. We say goodnight and head on our way. We'll check back in a couple of hours when we come back this way to make sure that they haven't reignited the fire.

In the bone-dry spring of 2018, as wildfires raged across Scotland, the Mar Lodge rangers put out seventy campfires in places which risked setting off a wildfire. At one of these sites, Kim put out the fire, only for the campers to reignite it, and for it to spread and set fire to a couple of small trees. The campers called the fire brigade and fled the scene. Fortunately, the site was only a few metres from the road, so it was promptly extinguished. In 2019, the extreme fire risk started earlier than ever, in late March, but the fire risk was dampened by a soggy snowy May. That didn't stop the rangers putting out or cleaning up after eighty campfire sites that year.

In 2014, a wildfire accidentally set off by campers ripped through the Luibeg woods, putting around twenty hectares of precious regeneration back to square one.[2] In 2003, another fire raged through eighty hectares of woodland and moorland in Glen Derry.[3]

There are arguments that a spot of burning can be good for the woods. It is the sort of 'disturbance event' that creates new ecological niches, and stops things getting too settled. There are even some beetle species that respond specifically to fire. Some people argue that creating firebreaks, ironically enough through burning, can stop fires spreading and getting too hot.[4] Studies from the Luibeg and Glen Derry fire sites suggest that when it comes to the Caledonian pinewoods, these practices are mostly distractions.[5][6] By far the most effective means of preventing wildfires in the pinewoods is to stop people starting fires in the first place.

Then there's the wood that people are putting on the campfires. Deadwood is a vital part of a woodland ecosystem. Something like 20 per cent of the species in Scotland are directly dependent on rotting deadwood for some part of their life cycle.[7] In some places at Mar Lodge it is all gone, used up by campers for their fires.

It's a frustration for Kim. 'The Scottish Outdoor Access Code is perfectly clear on this: if you've got a fire on peaty soil or in a woodland then you are behaving irresponsibly.' Kim is referring to the 2003 Land Reform Act, the landmark piece of legislation which gave everyone in Scotland legal access to most of the land. But this access comes at the perfectly reasonable price of responsibility.

It's a remarkably difficult job, telling people that they are doing something wrong. There are grey areas – what of the people who have been sensible enough to bring their own fuel, and light their fire on a shingle bank, right next to the river, far from the trees and the precious peat soil? What of the days when the midges are so bad that the only deterrent is a smoky fire?

'It's a judgement call, but fires make more fires. If people see a ring of stones, or a small patch of burned ground, they think it means that they are fine to light their own campfire.'

We carry on, through the estate, up to Derry Lodge, passing the ruined summer shielings of Glen Lui's previous inhabitants.

There are other problems. Overuse of 'honeypot' sites by campers can cause a build-up of human waste. One of the woods is known (by me at least) as 'Pooh corner'. Disturbance from walkers can cause black grouse leks to disperse, and we've seen how sensitive capercaillie can be to disturbance from people and their dogs. People trample precious woodland regeneration. Then there's the litter.

'The balance,' says Kim, 'is to allow as many people as possible to enjoy the woodlands without causing the destruction of what makes the woods special in the first place.' Finding that balance is a delicate, imperfect art. We walk around Derry Lodge, the grand, listed hunting lodge that sits at the foot of Carn Crom. It's a beautiful spot, surrounded by babbling burns and granny pines, and there's plenty of flat ground around the lodge which is suitable for camping. Derry Lodge was built as a remote, luxurious hunting outpost for the Earl of Fife and his guests. It was always a bit of a folly, however, and now it sits empty and somewhat out of place, like a person who arrives overdressed for a party. We scan across the glade, to Luibeg Cottage. This dilapidated little house is the last remnant of a long line of dwellings of successive generations of keepers, foresters and crofters. We are standing by the mountain-rescue hut. Just downstream is Bob Scott's Bothy, a well-used refuge for would-be mountain explorers. The buildings at this remote outpost of civilisation are a window into the changing relationship of people with the woods.

Kim shows me around the back of the lodge. It is surrounded by nettles. 'Nettles love fertile ground,' she says, eyebrows arched. 'And pinewoods are not naturally fertile.' We pull a bag of rubbish into the back of the pickup. It was left nicely tied up, tidily placed next to the lodge, but it was left here nevertheless. Rangers have to do regular litter picking here, four miles from the nearest public road. We discuss the lodge itself. It's a listed building, and the Trust has a legal obligation to keep it in good nick, but currently it is boarded up. One solution could be to turn the lodge into an eco-hostel, which is only accessible by hiking in. This might greatly reduce pressure from campers on the surrounding area. But it might simply increase the number of visitors to the area, creating ever more human pressure. This is beside the fact that people come here, as we have seen, to get away from it all, to rough it, if only temporarily. Increasing infrastructure might well be detrimental to the overall wild quality that people are searching for when they come to Mar Lodge. Then there is the tricky question of bin provision. 'We don't have bins, because we expect people to take their rubbish home with them.' Bins are expensive and time-consuming to maintain, and the evidence that bins actually reduce littering is mixed at best.[8]

There are new cultural phenomena to respond to. Instagram and other social media mean that places off the beaten track can quickly gain a cult following. The North Coast 500, the road around the far north of Scotland, for example, has proved to be a mixed blessing for the communities along the route. So far, Mar Lodge has remained (comparatively) off the radar, but it won't stay this way forever (especially if book sales are as good as I hope they will be).

There are several groups of campers down by the river. All of them are behaving, cooking on stoves. We share patter with a couple of groups, asking where they've been, where they're going. They are aware of the fire risk. They're happy to talk, for the most part. A few ask questions about the estate, the regeneration, where the best place is to see eagles. One group offers us a dram, which is regrettably declined. It's a gorgeous evening, and everyone is happy and smiling. 'The vast majority of people are very respectful and use the place sensitively. People come here because they love it here, and the last thing they want to do is trash it. And those that don't, well, most aren't being malicious, they just don't know any better.'

And so this is the ranger's job: talking, explaining, showing people how to enjoy the landscape without destroying it. It's frustrating, interesting, funny. A pinewood is not complete without its deer. Nor is it complete without its people.

The value of nature for our mental health has been well documented. A recent study of nearly 20,000 people found that participants who spent more than two hours a week in natural environments were significantly more likely to report good health than people who reported having no contact with nature.[9] Being in nature helps children hone their powers of concentration.[10] Time with nature is now prescribed by the NHS.[11] The wild is an antidote to the excesses of modern life, the stress, the pollution, the discontent, the emptiness, the FOMO, the creeping anxiety and guilt put upon us by the knowledge that our way of life is destroying the world. But it is much more than just an antidote, a flipside of the coin of civilisation. Mar Lodge is much more than Aberdeen's Arden. It is more even than a haven for wildlife, a store of carbon, an ecological experiment, an aesthetic amenity. For many thousands of people, the woods of Mar Lodge are an integral part of their lives. Their value lies in the richness of those people's experiences, their myriad stories and memories. Mar Lodge is more than a wilderness. It is living, breathing heritage, a web of the cumulative experiences of the people who visit the woods.

This, then, is the future for the wild woods of the Derry, the Quoich, the Mar Forest, of Luibeg, the Dubh Ghleann, the Beacan wood; the emerging forests of Clais Fhearnaig, the Upper Derry, Glen Lui; the trees marching up the slopes of Carn na Drochaide, Robbers' Copse, Carn Crom; the restructured plantations of Inverey, Compartment Two, the policies of the Mar Lodge grounds. These woods are a landscape of reconnection with nature, a reserve of hope for the future, an exemplar of what the future might hold for Scotland's depleted environment. By walking and exploring and sleeping in the woods, experiencing their beauty and diversity, we can root ourselves in that future. The value of these woods is in the inspiration they will bring to people to change their lifestyles, to demand a better, healthier landscape and better, healthier lives rooted in deep, complex and indeed contradictory relationships with the land itself. The woods have come a long way in the last twenty-five years, from a dying forest to a landscape of hope. These great woods are expanding, providing homes for more species, sequestering more carbon, fighting the good fight against climate change and environmental destruction. But in a world of 7 billion people, the true value of these woods is their ability to inspire people to live better, fuller, wilder lives.

Walking through the growing trees, I sometimes imagine coming back to Mar Lodge, decades in the future, after a long absence. I imagine Glen Lui full of willows, pines and birches, bursting with birdsong. I imagine stumbling into capercaillie in Glen Quoich, tripping over carpets of twinflower, finding wildcat tracks. The pleasure would lie not just in the inherent beauty of these interactions, but in the knowledge that I played a tiny part in its return to health. I would bask in its reflected glory.

This is the future relationship that we can all share with the environment – one where nature is not just somewhere to retreat to, but in the glory of whose joyous return we can all bask.

Welcome to the new great wood of Caledon. No fires, please, and take your rubbish home with you.

Part Two
ON THE MOORS

I've neglected the austere beauty of the Mar Lodge moors for too long. So I'll tell you about sunrise from Cnapan nan Clach, a forgotten hump of a hill on the way to An Sgarsoch.

The reason for being at such a remote place at sunrise is, of course, birds. Another early start, another existential crisis, this time in the name of the British Trust for Ornithology's excellent, vast, volunteer-led national Breeding Bird Survey.[1] They're very strict about starting the survey early in the morning. When the survey area is an hour's hoof from the end of the vehicle track, which is the best part of an hour's drive from the Lodge, that means a very early morning indeed.

I park the Land Rover at Bynack Lodge, a venerable ruin, and the sky is turning from starry black to bruised blue. Here are the remnants of another old hunting lodge, long occupied, the former home of the last native Gaelic speaker in the glen, derelict after suffering a fire in the 1950s. The hearth is still here; people still make campfires around it, though obviously the Trust doesn't encourage fires out here. There are a few larches and a grand old sycamore. The old lawns and garden still grow lawn and garden plants, yarrow, daisies, remnants of past human intervention. Bynack Lodge possesses the melancholy grandeur of heritage recently lost, its aura post-apocalyptic. I walk out through the bog, serenaded by curlews. Somewhere around here is the bubbling twirking of a black grouse lek.

I squelch on through soft boggy ground, all moss and puddles and cottongrass, with heather on the drier hummocks. After a few hundred yards I put up a brace of red grouse. Further up, I skirt around a group of stags, trying not to send them into flight. They don't look their best at this time of year. Their antlers are velvety or missing, and their coats are lacking their beautiful russet autumnal hues, but it's always nice to get a close encounter with a wild beast. Clammy, breathless and sweating despite the cool air, I hit the ridge and can't help taking a detour to the summit of Cnapan nan Clach, cloudberry growing out of the heather at my feet. The sun rises to the east, over Glen Dee, and I'm bathed in the sort of warm light that is usually only found on Instagram. The ground is transformed from monochrome to glorious red. Down on the *bealach*, in the usual spot, a golden plover pipes its lonely, rusty-bike-wheel call. Below lies the Bynack Burn, as remote a place as you'll find in Scotland, rarely walked through except by the occasional deer stalker or grouse shooter, or me, twice a year, for bird surveys. This is a glen of peat hags and wet flushes, a web of ancient pine stumps poking through the peat, the last vestiges of an ancient forest, lost to climate change several thousand years ago. In the distance, Beinn Bhrotain hunches over the western Cairngorms, Ben Macdui sits a bit more upright, and Beinn a' Bhùird marks the other side of the estate. To the south-west is Glen Tilt and the Atholl forest, to the north and west Glen Geldie, leading to the massive, regenerating pinewoods of Glenfeshie, and thence Speyside.

Such light, warm beauty is rare on these wet hills, and it is reason enough to put off the survey for a few minutes. The beauty of the great moors and bogs is at least in part a masochistic one. It is felt in the discomfort of wet feet and in the mizzle which creeps into your bones and the aching limbs of long walks. Indeed, it is this dampness, the ache, which imparts the loneliness and nostalgia of the place, its slight, melancholy spookiness, its, I'll say it, despite the fact that I know it to be untrue, emptiness.

The word 'moorland' goes back to Old English. The Gaelic equivalent *mòine* is perhaps better translated as 'moss', as in the Mòine

Mhòr, the big moss, which sits high up on the shoulder of Cairn Toul. The moorlands of Mar Lodge comprise the Mòine Bhealaidh, high up above 3,000 feet, which beckons in Munro-baggers and relieves them of their shoes; the Duke's Moss, an ancient blanket bog of the Geldie; the dry heath of Dalvorar; and the deforested moors of Glen Geusachan and the Upper Derry, which were woodland not so long ago. Some of the moors of Mar Lodge have been largely treeless since the Ice Age, others for thousands of years, others for hundreds. Some of them are 'natural' landscapes – largely treeless through natural forces, like the montane heaths and the blanket bogs. Others are more 'cultural landscapes', made treeless through human actions, and kept that way by centuries of deliberate management.

The word 'moorland', then, covers a multitude of different habitats, social constructions and sins. Just as 'woodland' might mean ancient Caledonian pinewood, Atlantic oakwoods or plantations of Sitka spruce, so 'moorland' might equally apply to the treeless deer forests, wet blanket bogs, heavily managed driven grouse moors or high mountain heath. Over the years, as landscapes have slowly changed and language has struggled to keep up, 'moorland' has come to simply signify a treeless landscape. In this linguistic shift are echoes of the history of another word, 'forest'. Scotland's deer forests are famously treeless – they are moors.

This etymological vagueness is unfortunate, because there's trouble brewing on the moors, and lack of clarity in words leads to lack of clarity in thought. In fact, there is no more bitterly contested Scottish landscape than its open uplands.

The majority of Scotland's moors are managed principally for Highland sport. Andy Wightman's forensic exploration of Scotland's land, *The Poor Had No Lawyers: Who Owns Scotland*, found that in 2002, 4.5 million acres of land, or 23 per cent of the whole of Scotland, was managed primarily for sport shooting.[2] This means that the cultural and economic act of sport shooting has a huge impact, for better or worse, on an enormous area of land, the majority of which can be thought of as some form of moorland. It is these impacts, and the often complicated relationship between sport shooting and environmental conservation, which we'll mostly be looking at in this section.

First, a brief overview of the alleged ecological pros and cons of Highland sport and sporting estates. For sports folk and their supporters, the moors are a haven for wildlife, long cared for by generations of sympathetic landowners and their gamekeepers. They are a landscape which, without care, would revert to a scrubby wasteland and be the worse for it.[3] Proponents of Highland sport and sporting management argue that red grouse shooting, stalking red deer and fishing for salmon provide jobs and income to vulnerable, rural communities.[4] Moorland management techniques such as muirburn (the ancient practice of burning land on rotation to promote heather growth) and legal predator control[5] (killing foxes, weasels, stoats and crows, for example) increase populations of both game species like red grouse and black grouse,[6] and also declining species like curlew and mountain hares.[7] They argue that the open heather moors of the east and the grassy deer forests of the west are the way that the landscape should be, and that Highland sporting culture protects these landscapes from change and destruction, and that without the influence of sports folk, our moors and the creatures they support will be lost to a wave of non-native spruce plantations.[8]

For their opponents, the moors are places of excess and snobbery, a post-Clearances landscape managed for the benefit of the lucky few rich and well-connected enough to have a punt at shooting a few dozen brace of grouse, stalk a stag or two, or fish for salmon.[9,10] They argue that the moors are private fiefdoms, run with almost no public oversight and owned by influential, privileged elites, where new hill tracks are built with almost no oversight[11] and leave near-permanent scars across the landscape. These are places where illegal bird of prey persecution continues unabated, sixty years after legislation was first put in place protecting the birds.[12] Where thousands of snares and traps are legally set, catching hundreds of thousands of vermin annually,[13] alongside non-target species including dippers,[14] ring ouzels,[15] merlins,[16] common gulls,[17] pine martens[18] and even golden eagles.[19] Where the use of muirburn has intensified in recent years, and spread to blanket bog areas,[20] drying out soils and peat, increasing flood risks,[21] exacerbating climate change[22] and intensifying rather than diminishing fire risk[23] and reducing biodiversity.[24] [25] Where the mountain hares which undoubtedly benefit from muirburn and predator control are subsequently shot in their tens of thousands to protect red grouse from the spread of ticks.[26] And we have already seen how divisive is the 'deer question'.

Sitting on the top of Cnapan nan Clach, enjoying the rare warmth of the early morning sun, listening to the golden plover, trying not to the disturb the stags, completely alone for miles in all directions, all in all highly content with my lot, I ask myself, how did we get here?

Let's retreat indoors, the lonely call of the golden plover still ringing in our ears, pour a dram of Royal Lochnagar and set the fire in the Mar Lodge library. At the mantel are fireguards stitched by Princess Louise herself, to keep her from reddening in the heat of the pine log fire. Above and around us are the stuffed heads of venerable stags, plaques indicating which venerable person they were lucky enough to have been shot by. Let's take a book from the shelves, and take a detour into the realms of sporting literature. We can do no better than the rip-roaring adventure-comedy *John Macnab*,[27] written, of course, by John Buchan, 1st Baron Tweedsmuir, the cabinet minister who also happened to be a novelist and a keen sportsman.

It's the early 1920s, and we join Sir Edward Leithen (ex-attorney general), Lord Charles Lamancha (cabinet secretary) and John Palliser-Yeates (noted banker to the Establishment), as they undertake a romantic, reputation-risking adventure to cure the ennui that comes with managing important affairs of state. Their real identities protected by the pseudonym 'John Macnab', and sheltered by wounded war-hero turned Conservative Party candidate Sir Archie Roylance, they issue a challenge to three estate owners in the Highlands. They will endeavour to poach a stag or salmon from each estate within a 48-hour window. The rules of the game are set, the challenge is accepted by each estate, and adventure ensues.

We are reading this ripping yarn because literature is a window into the soul of the writer and their culture, and *John Macnab* has been taken to heart by sports folk in a way that is unparalleled in the canon. As is regularly the case, life has imitated art, and now 'bagging a Macnab', albeit a legal one, is a highly respected test of skill in sporting culture. No longer poaching, the game is now to bag a grouse, a stag and a salmon in a single day. It's possible to book a Macnab at Mar Lodge, for those who are interested.

Buchan has the power to stir the heart of anyone who has found themselves entranced by the broad open reaches of the Highlands, by its crags and solitary corners, its lochans and bog pools, its clean air and its filthy weather. Sports folk can learn a lot from Buchan about their quarry and the Highland landscape. Buchan's knowledge of ecology and natural history is excellent. Sir Archie's fictional estate, the base of operations for John Macnab, functions as his archetypal Edwardian Highland sporting estate. 'You are to picture a long tilt of moorland running east and west, not a smooth lawn of heather but seamed with gullies and patched with bogs and thickets and crowned at the summit with a low line of rocks.' It is rugged and remote; this is a landscape which is not managed by burning and draining, but is a lively and diverse landscape of moor, bog and woodland. We are told that it is the 'most marvellous nestin' ground in Britain barrin' some of the Outer Islands . . . I've got the greenshank breedin' regularly and the red-throated diver, and half a dozen rare duck.' Sir Archie is more naturalist than sportsman ('scrabbling in the dell of a burn, he had observed both varieties of the filmy fern'), while his head stalker is a man of 'superior hill craft', though his 'chief recommendation was that he was a passionate naturalist, who was as eager to stalk a rare bird with a field glass as to lead a rifle up to deer'.

Buchan's Highlands, then, are not just a 'smooth lawn of heather', nor are they treeless. Indeed, the woods are of great importance to the sporting action of the novel. 'Great birch woods from both sides of the valley descended into the stream, thereby making the excellence of the home beat, for the woodland stag is a heavier beast than his brother of the high tops.' The Caledonian pines of a neighbouring estate, meanwhile, are an 'ancient remnant'. A peregrine nesting in the high corries is a source of pride rather than concern, a crowning jewel at the peak of the estate.

Buchan's knowledge of Highland sport is similarly expert; the book's descriptions of stalking deer are as exciting and as accurate as any you will find anywhere in literature. The rules of the wager at the heart of the novel are inextricably entwined with the rules and customs of good sport. Buchan describes in vivid detail the quintessence of the Highland sporting experience – the glorious discomfort, the skill, and respect for nature both game and otherwise. 'John Macnab' is not into sport for the size of the gamebag, but the quality of it. What is important to our heroes is not the taking of life, but of the way in which it is taken. Leithen's catching of a salmon on a poor piece of river, all the while sticking to using a fly-line (the gentleman's fishing tackle of choice), is a fine example of skill, subterfuge and sportsmanship. The stag killed by Lamancha in the final 'assault' is of course the finest of the estate's stags, not for the size and shape of its antlers, but because it is the wildest of the area's beasts, and has lived a long and good life. It fulfils the dual literary requirements of being both the 'best' stag in the forest and also being a suitable mark for the sporting cull.

Our ideal moor, then, is one of diversity, harbouring rare and endangered creatures, and cared for by Good Eggs. It is stalked by the quintessential sportsman, John Macnab, a gentleman in rogue's clothing. The quality of Highland sport is a 'great game'. Like Shakespeare's Arden it is somewhere that the rules of day-to-day life are suspended, and the higher rules of nature are given precedence.

So far, so good. Surrounded by nature, pitting ourselves against the elements and our quarry, the moor is a complex mosaic of woodlands and open ground which becomes a place where we discover our true selves and our rightful place in society. But look a little deeper, and Buchan's landscape hides the roots of the battle currently raging across our uplands.

In 1989, the critic Michael Young published a valuable deconstruction of the novel. He wrote that '*John Macnab*'s representation of class relationships in post-war British society is backward-looking and ultimately even repellent.'[28] This is hard to argue with. Buchan writes, 'It is a melancholy fact which exponents of democracy must face that, while all men may be on a level in the eyes of the State, they will continue in fact to be preposterously unequal.' Young notes that in *John Macnab* 'it is not economics and culture which subordinate the navvies to the gentlemen, but nature; thus the class hierarchy is simply an expression of the natural.'

For Buchan, then, the moors (and by extension, the Highlands) become places which are cared for by a benign dictatorship of the upper classes, to the benefit of nature and society, and without whom both society and nature would be lost. This is history written by the victors: there is no mention in the novel of the Clearances, or the land grabs which led to so many Highland sporting estates ending up in the hands of so few people. Glen Dee and Glen Lui, both once populated and cultivated, now lie empty of people, centuries of effort and life hidden away below the heather, the land potted with over 200 Scheduled Ancient Monuments. Michie, the chronicler of Deeside, wrote of the clearance of Glen Ey, which was once a part of Mar Lodge Estate, and the crofters 'compelled to give up their fields to the deer of the forest, to kindle in the breasts of those that remained a spirit of hatred against the offending proprieter.'[29]

There is no mention either of the species driven to the edge of extinction by keepers at the time he was writing – the Scottish wildcat, pine marten, white-tailed eagle, hen harrier, even ravens. Buchan's idea of a 'higher nature' trickles down into the practicalities of estate management in the novel. The estate owners, inherently competent and naturally Tory, look after the interests of their tenants as they do their deer herd, and are thanked for their service by being elevated to positions of high office. But one can't help but feel that, for Buchan, there is one rule for the elites and another for the plebs: when a cabinet secretary poaches a stag it is all fun and games, but were you or I to do it, there would be a prison sentence. The sketchily written, rustic locals provide an audience for the upper-class antics of the novel. By turns servile, lazy and cunning, they are happy with their lot, watching on with glee as their social betters shoot the game from the vast tracts of land to which they are denied access. The stalkers and gamekeepers may be respected for their skills on the hill, but they will always remain subservient to the landowners.

Buchan's moors become a place where sport shooting is elevated beyond entertainment, and even a necessary management tool, to a vital task for a functioning society in which people know their proper place. What we are seeing in this novel is the beginnings of the ideological tautology that lies at the heart of the problems of our over-managed moors: a moorland is healthy when it supports Highland sporting culture, and Highland sporting culture is necessary because it maintains healthy moors.

Life imitates art: *John MacNab* is a powerful example of how politics and history have profound impacts on both the nature of landscapes and the nature of our relationships with landscapes. Indeed, the historian James Hunter goes so far as to say (and with some authority), that Highland sport began as 'little other than a literary fantasy of what it means to live in the grand Highland style.'[30]

Moorlands in Scotland are bound up with identity politics to a greater extent than perhaps any other landscape in Britain. For some, heather-clad hills latticed with squares of burn marks and throbbing with far more red grouse than would naturally occur on them are a sign of well-managed land, a place of order, job opportunities and a temporary victory in the perpetual struggle against the encroachment of vermin, woodland and land reform. For others, those same moors are a cleared, blighted landscape, signifying only persecution, injustice and diminished biodiversity, a place of untapped ecological and economic potential and a stain on the soul of the nation.

There is no denying that Highland sport finds itself under a cloud. Recent intensification of upland management[*] has coincided with more people noticing these excesses, and successfully using social media and modern tech to highlight them to an increasingly large and increasingly receptive audience. The future of the sport was debated in the Commons in 2016. A new wave of private landowners, outsiders to the Highland sporting establishment, is balking against traditional moorland management, choosing instead to rewild large areas of the uplands previously managed as sporting estates.[31] In December 2019, a government-mandated review of grouse moor management recommended that, if there were no large-scale ecological improvements to the state of the moors within five years, then government intervention in the form of licensing sporting activities will be required.[32] The Scottish government immediately made noises that this timescale may not be quick enough, and that steps to license the sport may be fast-tracked, with or without environmental improvements.[33] Sport shooters, moorland owners and their powerful lobby groups argue that this will be the end of Scotland's moors. Many environmentalists argue that these potential reforms do not go nearly far enough. Either way, the moors are a changing landscape.

For all Buchan's questionable politics, his ecological education remains sound. Can landowners realise his vision of the ideal sporting estate, a mixture of birchwoods, pinewoods, deep thickets, salmon-filled burns, open bogs and moors? Can we retain the

---

[*] Of which I'm certain Buchan would not have approved – remember that back in his day there were fewer than half the number of red deer than is currently the case. Buchan was always one who felt that sport was only worth doing if it was in some way difficult.

benefits and beauty of the moors, their solemnness, wildness, wetness, their waders and grouse and deer, while also doing away with the worst excesses of grouse moor and deer management? Such is the idea, broadly speaking, behind the management of Mar Lodge's moorland zone, the huge area that makes up the south and west of the estate. Here, Highland sport remains an integral part of the estate's land management. Gamekeepers manage a (comparatively) small but healthy deer herd, and deer stalking with clients remains an important revenue stream for the estate. There is no intensive driven grouse shooting, no medication of grouse, no burning, no snares or traps, no peatland draining. Instead, the keepers offer days of walked-up grouse shooting, the original sport, with far fewer birds and more effort to shoot them. Populations of ground-nesting birds are helped along by a limited amount of targeted predator control, as they are on other nature reserves across the country. Peregrines perched on their crags are crowning jewels. And all this sits alongside expanding and regenerating woodlands.

The estate aspires to move on from the imperialist, problematic past of *John Macnab* and address the very real problems of the present. It aspires to emulate the things that make *John MacNab* a beguiling and rip-roaring read: the skill of the stalkers, their concern for and love of the land they stalk, their deep knowledge of and connection with nature and the landscape. The estate aspires also to collapse the idea of moorland and woodland as being mutually exclusive, treed and treeless landscapes, and instead treat them as nodes on a continuum of intact, healthy, connected habitats.

In this section are stories from people fighting for a new, revived vision of what Scotland's moors can be. We'll be finding wader nests and seeing what is predating them. We'll be searching for salmon from the high moorland burns to the open Atlantic. We'll be finding hen harriers and praying that they don't get shot on some lonely patch of grouse moor outside the estate boundaries. We'll be rewetting the moors, fighting climate change along the way. We'll be offering up a new vision for our moors – one which aspires to bring together both sides of an increasingly vicious debate. Moorland management at Mar Lodge has been deeply criticised by elements of the sporting community. As it turns out, the evolving management of Mar Lodge and other estates might just be a saviour of Highland sport.

7

Red Deer

In 1563 Queen Mary was present at a Highland hunt when, in the course of two months 'tenchel' driving, two thousand Scottish Highlanders collected in the wilds of Athole, Badenoch, Mar and Moray a huge herd of 'more than two thousand deer'. Many of these, owing to a sudden stampede of the herd, broke bounds and escaped, but notwithstanding, 'there were killed that day 360 deer.'

James Ritchie, The Influence of Man on Animal Life in Scotland, 1920

One of my best interactions with red deer happened by accident one late October. Returning from a longish walk, a friend and I came off a high pass and decided to take a shortcut down a steep, wet, rarely visited glen. Hemmed in by a waterfall on one side and a grassy bank on the other, unable to see beyond the steep gulley walls, we smelled the musty, goaty tang of deer, and the hairs crept up on my neck. Disturbing stags in autumn is a notoriously stupid thing to do. Another second and we could hear the clatter of antler on antler and the grunting as muscles flexed in effort. I peeked over a hump; an alarmingly large stag was seeing off a smaller counterpart, as a dozen or so hinds looked on. He snorted and I saw his breath.

We were far too close for comfort, of course. Stags are fidgety and unpredictable, and can be very dangerous in the autumn. Moreover, the gamekeepers would be rightly furious if they knew we were disturbing the rut. We tried to retreat, but not before the group caught wind of us. Normally this would cause them to flee. But the stag, blood surging with testosterone, held his ground. We tried to look big, and slowly walked away, resisting the urge to break rank and run for it. Another sniff of our human stink, his courage failed, he gave a desultory grunt and herded his harem off to safety.

Heart-in-mouth, wild stuff, and a good illustration of why red deer loom so large in our collective psyche. They are truly beautiful, remarkable creatures, one of our few remaining links with an older, more biodiverse Scotland.

The question, 'Where do I go to see the stags?' sits around number two in the list of the top five things I get asked while I'm out and about at Mar Lodge.* The problem is that it's an open question. It depends on the time of year, the time of day, the weather and the fitness and patience of the questioner. It depends whether they want to see them or photograph them, whether they have binoculars or not, how noisy they are, how many of them there are.

Anyway, here's the answer. Autumn is the best time of year to see red deer. After a summer of feeding up, the deer are in good condition. This is when they are looking at their best, their russet autumn coats thick and glossy, their antlers sharp and de-furred. Hinds are tending calves, but coming into heat, sending the stags into a frenzy of activity. Come on a fine crisp day, wait until late afternoon, and drive along the road from Braemar to the Linn of Dee. You'll hear the uncanny roaring of stags rattling around the glen. There'll probably be a rut underway on the Quoich Flats. If not, you'll see them coming out of the shelter of the woods around Inverey, heading to the open fields that flank this part of the River Dee, just as the light begins to fade.

---

* The other questions, in order, are 'Is the Lodge open today?', 'Which way is it to Derry Lodge?', 'What do you think about lynx and wolves, then?', and 'How do you pronounce "Quoich"?'

If you feel like throwing a little money the Trust's way, then you can join the popular stag rut watching events, which run in October. This is the responsible way to enjoy the spectacle of red deer; you can be assured that you won't be disturbing them (as we foolishly did), but you'll get good views and hopefully learn a lot too. If you want to throw the Trust significantly more money, you can sign up not just to see a stag, but also to shoot one.

Another place to see stags is the Stag Ballroom. There's plenty in there.

It's a curious building. From the outside it looks like a Swiss hunting lodge, or just some sort of fancy oversize garden shed. It's a large, rectangular timber construction, clad in wood painted white, with a steep, red-tile roof. About its cladding is another layer of thin diagonal trellis, painted bright red, the sort of thing you would grow honeysuckle up in your garden. It has bright-red timber doors and snow-catchers on the roof (for insulation). There are no windows, only skylights. It looks like a Swiss hunting lodge because that's what it is meant to look like – Swiss hunting lodges were very 'in' at the time it was built.

The ballroom looks rather out of keeping with the rest of the buildings and grounds; the Lodge itself is a large, imposing pink granite affair, again in the Swiss hunting style, but even more oversize than the ballroom. The ballroom sits somewhat demurely at its side, tucked out of the way. The reason for this is that the ballroom is older than the Lodge. It was originally on the other side of the glen, an addendum to the second Mar Lodge, cutely called Corriemulzie Cottage, but this burned down in 1895.[†] So taken was the Duke of

---

[†] We are currently on the third-and-a-half iteration of Mar Lodge. The first, Dalmore House, was flooded, and the second burned down. The third iteration was constructed in 1895–98, with Queen Victoria laying the first stone. In 1993 the Lodge was significantly damaged by yet another fire and had to be substantially reconstructed. In 2016, it very nearly flooded again but got away with it by a whisker.

Fife by his ballroom that he ordered it be dismantled, piece by piece, and transplanted to its current spot, next to the third-and-a-half Mar Lodge. It has been in its current location since 1898.

The ballroom was constructed to accommodate the ghillies' balls, which are still held, once a year, after the Braemar Gathering. Grand ceilidhs would be thrown for the estate staff, the ones remaining after the Clearances from Glen Lui and Glen Ey, to keep them happy in their work. A golf course was constructed for a similar purpose across the Dee, though this is now long gone.

Of course, the social protocol of the time demanded segregation between social classes. Not that the great and the good were above visiting the ballroom. Far from it. Recessed into the wall is the royal viewing area, from which Queen Victoria and other notables of the Balmorality would watch on as their servants and ghillies ate, drank and made merry. It must have made a fine scene. The locals, humbly enjoying their rustic traditions, the band playing, the whisky flowing. And above and around the merry dancers were hanging the trophies of hunting exploits, the skulls of stags, a noble quarry reserved for the noblest of people, each skull the embodiment of a fine adventure on the hills. Looking down from every conceivable piece of wall, truss and ceiling, 2,435 skulls were watching the merriment unfold below them. That's right. There are 2,435 stag skulls in the Stag Ballroom, all of them shot on Mar Lodge land in the late nineteenth century, preserved for posterity.

In this hyper-real, desensitised world of ours, where meat comes wrapped in plastic and computer games replicate war zones for our amusement, there are few things that remind us of the true viscerality of life and death. The Stag Ballroom is one of those things. It is a royal folly, a political statement, an unashamedly, triumphantly elitist expression of the love of a particular type of hunting. It is a celebration of the deeply flawed understanding of the Highlands as a place *foris*, outside the normal conventions of modern life. The cult of the red deer has influenced everything in the Highlands, even the architecture. The Stag Ballroom illustrates the totemic nature of the stag in Highland culture. It explains the deep-seated opposition of the landed gentry and their acolytes and copycats to any possible threat to the Highland sporting way of life. The Stag Ballroom is a particularly potent expression of cultural hegemony, of the oppression of both land and people for the benefit and the sport of the very few. As we saw with *John MacNab*, land management and culture are two sides of the same coin. Highland stalking was a statement of political will; this is our land, the ballroom says, these stags are the fruits of its fecundity, brought to rightful dominion by our skilled hands, and we are generous enough to bestow on you a place in which to dance. The Stag Ballroom is a grisly theatre of pomp and power, a statement of oppression over people and the environment, and an elitist expression of the Natural Way of Things.

But that was then, and this is now. These are the cold, hard, economic facts of the deer-stalking industry. In 2016, according to the sporting industry's own figures, the total direct expenditure on deer stalking in Scotland was £49 million per year, and the entire sector directly supported 840 full-time-equivalent jobs.[1] These figures are extraordinarily low considering that a huge area of Scotland's uplands are devoted primarily to the production of red deer for stalking. High deer densities maintained by many stalking estates also take their toll on the public purse through subsidies to fencing, damage to agriculture and forestry, and road accidents. In 2016, SNH reported that 'evidence gathered to date suggests that management of deer in Scotland results in a net monetary loss for both the private and public sectors.'[2] Such low economic performance might be acceptable, were there a strong ecological case for maintaining deer numbers at such historically high figures. But there isn't. An overabundance of deer may be desirable for deer stalking, and indeed drives up the value of sporting land, but it leads to environmental destruction.

In 2013, the total land area owned or managed by environmental charities in Scotland (with National Trust for Scotland being the largest landowner of these) was just 8 per cent of the size of the land area of private deer-stalking estates.[3] A report by the University of the Highlands and Islands' Centre for Mountain Studies that same year found that landowning environmental NGOs employed 305 full-time-equivalent staff directly on their land – 36 per cent of the number of people directly employed by the entire commercial deer-stalking sector and five times more jobs per square kilometre, not including the HQ office workers supporting them.[4] The report found that landowning environmental charities directly spent £37 million per annum on the sites they manage – 86 per cent of the amount directly spent by deer-stalking estates, and ten times more per square kilometre.

Mar Lodge, a Highland sporting estate owned by an environmental charity for the benefit of the nation, sits between these two ways of managing the Highlands, and provides an intriguing potential model for the way forward for Highland stalking. It is argued by those in favour of retaining the status quo that reducing Scotland's deer herd will inflict economic harm to sensitive rural economies. The Mar Lodge story suggests that this might not always be the case: the number of stalkers on the estate has remained broadly the same as in the commercial sporting days. Highland sport on the estate pays for five full-time gamekeepers, plus a couple of ghillies in the stalking season. It brings in enough to pay for itself, plus all of the woodland restoration work that the stalkers keep themselves (extremely) busy with. But Mar Lodge is a diverse estate – gamekeepers account for around a quarter of the estate's staff and generate around a quarter of the estate's income. It is often argued by sports folk that environmentalists hate deer. Again, the Mar Lodge story suggests otherwise: there are over 1,500 red deer on the estate, more than enough to ensure the long-term health of the herd, and also enough to link up deer populations across the Cairngorms and Grampians.

It is true to say that the population density of red deer on the moorland zone at Mar Lodge is far lower than can be found on other sporting estates, but it is still more than enough to ensure that stag stalking is at capacity for what the keepers can offer for sporting guests. In fact, the comparatively low deer density found in the sporting zone at Mar Lodge (8–10 deer per square kilometre rather than the 15–25 deer per square kilometre more commonly found on deer forests)[5] is not only driving down instances of ecological damage caused by an overabundance of deer but is also driving up the quality of the deer themselves. A study published in 2019 looked at the body weight and fecundity of over 36,000 deer harvested from nine estates, including Mar Lodge, over a 35-year period. It showed that the weight and fecundity of deer increased when deer numbers were reduced.[6] Fewer deer means better quality and more fertile deer. This means that, to maintain a lower number of deer on the ground, you still have to shoot a similar amount of deer to what you would be shooting to maintain a larger, less healthy herd. Having fewer, better deer also drives up the quality of the hunting itself. Stalkers and their guests are required to put in more effort for their prize, which, as we know, makes the prize all the sweeter. While the sporting cull focuses on the older, infirm stags, better food availability (with fewer deer there is more vegetation to go around) drives up the overall quality of the stags, meaning that those who want a trophy for their walls are more likely to have a larger one.

The Mar Lodge model for stalking is simple. Maintaining a comparatively small herd of higher quality deer offers a higher quality sporting experience in a healthier ecosystem, with no impact on financial income and with significant ecological benefits across the moors, bogs and woodlands of Mar Lodge. Deer stalking at Mar Lodge is still done in the traditional Highland way – a skilled stalker takes a guest up onto the hills and bogs, a single rifle is used, and deer are killed with a single shot. The Mar Lodge tweeds are still proudly worn, local whisky is still toasted at the end of the day. These are not just markers of proud cultural convention, but also a unique selling point. People come from across the world to perform Highland sport. Every year the Mar Lodge stalkers welcome

American, German, French and Scandinavian hunters looking to try something a little different to what they can experience on their own beat. The model works: deer stalking at Mar Lodge is fully booked a year in advance.

But the really revolutionary change to deer management at Mar Lodge is that, for the first time in its thousand-year history, sporting success is measured in benefits for the nation, rather than benefits for the landowner.

The ballroom is not just a theatre of imperialist oppression. It is also a highly instructive place for the student of ecology.

The royal viewing area, recessed from the main ceilidh area, now serves as the bar. Around its dark walls are hung all the heads which weren't quite up to scratch, the curios and oddballs that don't deserve to be displayed with the others. Here is a head with a perfectly straight set of antlers, like an antelope. Here is a single antler with the other broken off, presumably mid-rut. Another antler, damaged while it was growing, has flopped sideways, and looks like it is melting in the heat of the ceilidh. Here is a 'switch', a deer with antlers that have no points coming off them. Stalkers get rid of these beasts, as their antlers are particularly dangerous during the rut. One skull presents an 'Atholl head', an oddly square arrangement of the upmost antler points, which is a distinctive feature of stags from the truly vast Atholl estate. Another, huge skull has no antlers at all. This is a hummel, a stag which doesn't grow antlers. Antlers are bone; they take a huge amount of energy to grow. Hummels instead put all of this energy towards body growth, meaning that they can reach huge sizes. Stalkers dislike hummels because they are capable of seeing off any normal deer at the rut, owing to the counter-intuitive fact that stags 'square off' by way of body weight rather than antler size. This means that hummels can dominate a breeding season. In the 2019 sporting season Dom (from Chapter Three) told me of a hummel on his beat, his Moby Dick. Over the course of the season he got to know its haunts and movements, but it never presented a shot, and escaped for the year. We talked about potential MSc theses for him on the nature of hummels.

Several dozen taxidermied heads are arranged around the ballroom, below the ranks of the skulls, just above head height for the average ceilidh dancer. Curiously, many of the antlers on these heads belong to the same stag, a beast from Windsor Great Park, another ancient royal hunting forest set apart a thousand years ago for the gentry to use as a playground. They have been transposed onto other deer heads (some of which come from hinds, a joke on the part of the stuffer). The first thing to notice about these heads is that the antlers put the lie to the myth that you can tell a deer's age by how many points there are on his antlers. Instead, you can see a gentle improvement in the quality, size and number of points on the antlers as the stag reaches maturity, and then a gentle decline as the beast 'goes over' and becomes a candidate for the annual cull. The Windsor stag who donated his antlers to the ballroom was a long-lived beast, still proffering antlers at sixteen years old, an age far beyond what most hill stags will reach.

Take another look at those taxidermied heads, and you'll see something else peculiar. Those antlers are significantly larger than the antlers on the Mar Lodge skulls. Some may be twice the weight of the antlers on the skulls. These fine specimens lived a cosy and cosseted life in the warmth and woods of Windsor Great Park. As we have seen, we think of deer as a species of the open hill, but they are actually far more comfortable in the woods and even wetlands. The moors of the Cairngorms, cold, wet, lacking in cover and with little but heather to eat, are a poor place for growing large red deer.[7] Give them the pinewoods and the birchwoods and they'll grow far larger and give you a much more impressive head. I've a couple of shod antlers in my house to prove it – two identical antlers, one found on the moor, one in the birchwoods. Identical, except for the size. The birchwood antler is twice as heavy.

One of these stuffed heads has a placard around its neck. It was shot by the Duke of York in 1922, at Black Bridge in Glen Lui. You'll note its size and note how much larger it is than the other Mar Lodge stags. It has fourteen points, a monarch of the glen. But if you look at its improbable antler growth, you'll wonder to yourself, is this a ringer? Did the keepers stick some park deer out in front of the duke and tell him it was the finest stag in the forest?

The time for ringers, for the false glory of canned hunting, is over. For all the history, the tweed, the tartan, the hunting pursued at Mar Lodge these days strives towards a symbiosis with nature. The ultimate aim for the estate is, indeed, a symbiosis between environmental conservation, Highland sport and open access for all. Highland stalking has always been a cultural gesture. At its worst, it seeks to control, to arrest development for the sake of tradition, to derive pleasure from mastery over the environment to the detriment of the Highlands' humans and environment. At its best, it aspires to improve the lot of the people, the deer and the environment of the Highlands, to increase biodiversity, provide jobs and food and become part of a thriving, equitable rural culture and economy.

In most gamekeepers I have known, there is an instinctive tug towards the best in stalking, and a deep love and knowledge of the creatures that they hunt. But Highland stalking sits in a precarious position, challenged like never before and regularly called upon to defend its largesse, inequality, privilege and economics. In January 2020, a major government report called for a large reduction in the number of red deer in the Highlands, with an upper limit of ten deer per square kilometre and sweeping changes to deer management.[8] The Scottish Gamekeepers Association, the ones who accused Mar Lodge of 'abuse' of deer back in 2010, described the report as a 'hammer blow'.[9] Long-held truisms of deer ecology, biology and management are being tested by climate change. To survive, Highland sport will need to adapt, to shed what doesn't work and keep what does. The Mar Lodge stalkers are demonstrating what may well be the 'new normal' for the deer forests of the near future. Other hunting models are being offered up: some suggest community hunting rights, opening Highland sport to poorer local residents and allowing them to shoot for the pot, should they wish to. Others suggest quotas for game based on biological benchmarks like the average weight of shot deer, a proxy for the overall health of the deer herd. Meanwhile, wealthy landowners like Paul Lister, the Rausings and Anders Povlsen are doing away with the old, holy trinity of deer, grouse and salmon altogether, bringing a whole new culture to the Highlands, and a significant amount of wildlife with them.[10]

The Stag Ballroom is no longer just a theatre of power and privilege. It has taken on a new life as a wedding venue and events space, open to all. It sits, sometimes uneasily, as a bridge between the Highland sporting culture of old and a new sporting culture being forged in the fires of the Anthropocene. It is, in its own idiosyncratic way, the temple that the red deer deserves.

8

Sphagnum Moss

It takes a bit of practice to love some places. Carn an Fhidhleir (Carn Ealar, hill of the fiddler) and its environs are just such a place. The good folk of the Walkhighlands online community have decided between them that the hill deserves a mere 2.3 stars out of 5, and it is ranked 276th out of the 282 Munros.[1] For comparison, Sgurr nan Gillean, the finest of the Skye Cuillin, sits at the top of the rankings with 4.7 stars. The highest-ranked Munro on the estate is Braeriach, sitting at 44th, with 4.11. It seems that the members of the Walkhighlands community skew towards the pointy hills of the West Highlands, rather than the rounder delights of the Cairngorms.

Shows what they know. Carn an Fhidhleir is wild land without pretension. It turns its back on egotistic adventurers. Yes, it may regularly be a midgy, cleggy hellscape. And yes, when the rain comes, which it does with extreme regularity, there is no shelter for miles. But it is also a place of intense beauty, with huge skies, subtle colours and rare wildlife.

At first glance it appears barren and inhospitable, but not in the good way. There are no grand, rocky corries, no heart-stopping cornices, no razor-thin ridge lines. Unless you are a deer stalker or a Munro bagger or happen to need some serious alone time, there is very little reason to go there. Walking up to it through the bealach with An Sgarsoch is a pain. The burns weave about in an unhelpful fashion, cutting their way through metres of peat, forming hags, deep black gullies into which you sink up to your knees. There are few paths. To get to the top, you'll need to bogtrot. Most people give up on the place after the first glance.

At second glance, it is still an unloved place. But then the visitor may begin to be impressed by the remoteness of its aspect, and grudgingly admire the sheer, bloody-minded distance it has parked itself from everywhere else. The peat hags are remarkable things and have their own stubborn character. A second glance and you'll spot a deer or two, or some grouse, and think that it's not *completely* desolate. You may by now have worked out which shades of green in the bog will support your weight and which shades will collapse under you, leaving you stuck up to your knees in stinking water.

At third glance you begin to see the human and ecological stories that are imprinted on the landscape. Those hags finally start to register in the mind. At their base sit the perfectly preserved root plates of pines, thousands of years old. It may begin to dawn on you that this place wasn't always treeless, that it was once a verdant forest. A bit of research and you'll discover that those trees died thousands of years ago, swallowed up by rising peat, fuelled by a warmer and wetter climate, in all probability exacerbated by forest-clearing humans. Carn an Fhidhleir, the hill of the fiddler; there must be a story there. Its neighbour, An Sgarsoch, used to be a cattle market. This is like the Irish bogs of Heaney's poems – the bogs that reveal and hide the touch of humanity, unnerving and living places, the layers of peat haunted by what has come before.[2]

It may take you four whole glances before you really even notice the wildlife. The creatures that inhabit this open landscape are remarkably difficult to find. A common lizard breaks cover, an adder basks in an open patch on a sunny day. An eagle flies over a mile or two away, a short-eared owl hunts for voles. The snipe, the dunlin, the golden plovers keep unsociable hours, doing little during the middle of the day, which is when you'll probably be there, because it's a long trek to get into the area. All those pools of water you've spent hours trying to avoid are actually great for dunlin and golden plovers. Meadow pipits are plentiful and feed the merlin which has been there all the time, but you still haven't noticed somehow. Skylarks sing their guts out on the drier, grassier bits. There are goosanders and dippers and common sandpipers in the river, and pied wagtails, far away from their adopted urban haunts, wagging away in their 'natural' habitat. Sand martins have made burrows in the burnside. You stumble over mountain-hare poo. A vole may jump into your lap while you eat your piece, as happened to me one lunchtime out that way.

It will probably take your average punter about five glances to notice what makes this place really special; what ties together all those disparate things you've been noticing. This landscape doesn't make sense unless you happen to notice the sphagnum moss.

May 2019, and we're dodging the showers in the shadow of Carn an Fhidhleir in a beaten-up estate Land Rover. In the front, Skitts and myself. In the back sits Indy, a very sweet and very handsome Golden Labrador, with a knack for mooching people's lunches. There's a fair idea that the skill of the fieldworker is inversely proportional to the state of their clothing. Skitts goes surveying in jeans and a T-shirt. Every year, at the beginning of the season, he buys a single pair of wellies, the cheapest ones you can buy at B&Q, and they see him through the field season. For some reason, Skitts reminds me of Brian Cox. Today, we're looking for dunlin.

The Geldie is a big wide flattish valley, sprawled out between Carn an Fhidhleir and An Sgarsoch to the south, and Beinn Bhrotain to the north (pronounced 'vrotten', as in 'rotten Bhrotain', as Skitts calls it, remembering his ghillying days). It looks fairly benign, but we are deceptively high up here. I check the GPS; we're 570 metres above sea level. We set off into the bog. Carn an Fhidhleir leers above us with the air of a mountain which has seen it all and doesn't much care to see any more.

Like most people (or so it seems), Skitts worked at Mar Lodge as a ghillie for a couple of years in the seventies. He remembers that they weren't allowed to take ponies through the bog we are working, as it was too wet and dangerous. He points out an old peregrine nesting site. It wasn't ideal for the birds, so he fashioned a platform and worked it into the rock, and peregrines used it for a decade or so. He talks of finding goshawk nests, redwing nests, proper rare stuff. He's big on bird ringing – the scientific practice of putting small rings around birds' legs to identify individual birds in the field. He likes golden eagles and waders best, but he studies anything he can find. Every year he heads up to the far north of Norway to study waders – purple sandpipers, broad-billed sandpipers, the exotic stuff.

Somehow, he managed to fit all of this in whilst also running an electrical component supply business. But this year, he's retired.

We disperse out across the bog. The rain is thinning, but the wind is putting us at the edge of what we can deal with. As is often the way with serious bird monitoring, it's fine if you can't see the birds, really. But if you can't hear them, then you are in trouble.

This is a newish monitoring site. It's quite big, so that we can get enough data, but small enough that it can be comfortably studied in a day. The idea is to get to know the area extremely well over a few seasons, understanding what makes it tick. Our patch is flattish, undrained, and very wet. Heather does not survive in abundance here; it creeps through the high ground, surrounded by cottongrass, the occasional tuft of deergrass, and lots of sphagnum. In other words, it is perfect for waders.

After ten minutes, I hear a dunlin alarm calling, and I pick it up flying left over the pools and sphagnum and low heather and deergrass. Skitts is already on it. As is Indy. She's on point, frozen, nose down in a tuft of vegetation. We're in. 'Got a nest!'

I walk over, careful of my footing, not wanting to hear either a squelch as I fall in the bog or a crack as I tread on an egg. It takes me a while. I know where it is but I can't see it. Then, like a magic-eye picture, it appears. Four tiny, dark speckled eggs in a cupped nest, hollowed out of a small patch of deergrass. The eggs are perfectly camouflaged: grass-green with swirls and spots of darker colours matching the shadows, making them difficult to notice with a casual glance.

Indy gets a dog treat. 'It can take hours of watching to find a dunlin nest without her,' says Skitts. This summer she'll go on to find thirty or so willow warbler nests, as part of a long-term study Skitts is doing on the species. The nest keeps going in and out of focus. It is too well camouflaged for me to see, even when I'm looking directly at it.

Dunlin are truly excellent creatures in ways that few other creatures are. They are small waders, about the size of a starling. They can live comfortably to over a decade old. In winter they are fairly drab, brown and white. In summer, they are among our most attractive birds. A bold black belly bordered white, mottled chestnut back, smart bill. They are true arctic birds with a circumpolar distribution. The British population is the lowest latitude population in the world.

In Scotland, they breed from sea level to well above the 900-metre contour. They breed in the machair and marshes of the Western Isles, and in the vast blanket bogs of the far north. They also breed on select moors, where there is plenty of water and bog pools. They breed in the tundra of the Cairngorms. And most of the time, most people would never know that they were there.

'How long do we have before the eggs chill?'

'Oh, a good half hour. Remember, these are arctic birds – they can deal with it a lot colder than this.' I take pictures of the location, and a GPS reading, and pause to admire it. Skitts is slightly surprised to find the nest, and it seems fairly well on. 'It's all about timing. This year the late snow has put things back. We'll have to see what else we can find.'

The showers aren't a problem, except that being wet makes me grumpy. I look around; we're in the middle of some prime dunlin real estate. There's lots of cottongrass, far more cottongrass than there is heather. There are small pools of water, not lochans, more puddles, but they lattice and intermingle. This is a distinctly wet place. And everywhere there is sphagnum moss, four or five species of it at a time. Water breeds insects, and insects feed waders. Dunlin know a good bog when they see one. Skitts takes a small pad of moss and sticks it on a hummock a couple of metres from the nest. In time, it will bleach in the sun, leading us back to the nest.

The adult is still in sight, giving a melancholy chirrup from time to time, letting us know where it is, and perhaps could we hurry along please. Beyond the extraordinary camouflage, dunlin nests are extremely hard to find because the adults are masters of misdirection. When a potential predator comes along, they skulk off their nest, scurrying below the vegetation, out of sight. Then they make a big show of being terribly alarmed and fly off. It's a double trick – either you follow the bird to where it flies off to, or you think that the nest is where it flew off from. Either way, you are wrong.

The study site came about in response to a few things, but mostly the fact that I'm interested in waders. There's never enough time to monitor everything on the estate, but I make time for waders. They are having a rough time of it at the minute, so the more we can know, the better. There are also potentially land management implications to our study. The stalkers are currently reducing the amount of deer in this part of the Geldie, and we want to know what will happen to the wader populations when deer numbers go down. We're crunching through dwarf birch, a diminutive, subarctic relative of the birch trees you find across the country. I find tendrils half a metre long and a couple of centimetres above the height of the heather. That couple of centimetres is extremely interesting, as we'll see in the next chapter. By building up a detailed picture of nesting density and breeding success, we can see how changes to the site, be they ecological or climatological (or both), affect our waders.

After four hours we've picked up eleven dunlin territories or nesting sites and four golden plover territories. It's encouraging – breeding dunlin aren't exactly ten a penny in Scotland, and they're still recovering from substantial declines in the 1980s and 1990s.[3] Breeding success has not been great at our site for the last couple of years – torrential downpours just when the chicks are hatching can completely wipe out whole generations of birds, while drought conditions can ruin food abundance at critical times. This is not catastrophic in the long run, as dunlin are long-lived birds and high chick mortality is built into their breeding strategy. But as climate change bites and extreme weather becomes more likely, dunlin colonies will become more vulnerable. Large, healthy populations spread out across large areas will become the key to long-term dunlin survival in Britain. And on the moors, those populations are dependent on sphagnum.

The rain really sets in, and it's time to be off. On the way back we spot an osprey. They regularly fish here, but are yet to breed on the estate, despite the wealth of suitable nesting sites. This one has come from Speyside – they'll travel a long way for fish. Skitts says, 'I once saw one along the Gairn road, carrying a fish from Deeside. It was following the main road, so I followed it. It crossed over into Donside, and then went up to Corgarff, and then over into Speyside. After twenty miles I lost it, still flying.'

Sphagnum is certainly odd-looking, but it has its charms. Deep greens, yellows, browns and reds, sphagnum tends to grow in patches, clumps or hummocks, depending on the species. Each individual plant is capped by a rosette of densely packed leaves. This cap gives way to a long, squishy stem, which gives each individual plant a shape something like a jellyfish. The closer you look, the richer and more subtle the hues and lines become. Look at them for long enough, and you'll see that sphagnum hummocks are a pointillist feast for the eyes.

Sphagnum is one of those geeky things beloved by a very small, very specific group of people, and completely ignored by everyone else. At least, that was the case until very recently, when suddenly conservationists realised that it may well be Scotland's most important plant. Sphagnum, you see, has almost miraculous qualities. Sphagnum is the keystone plant of some of the UK's most important and (until recently) least-bothered-about habitats: bogs and peatlands. And there is a swingeing renaissance occurring in peatland conservation in the UK.

Throughout history our peatlands have been widely thought of as a wasteland in need of improvement. Over the centuries they have been drained and burned, the peat used for fuel and to fill our gardens. In Deeside there is a long history of locals using peat cuttings on the hills for fuel. In more recent years our peatlands have been planted up with non-native trees. Humans have tried every conceivable mechanism to make the bogs make money. None of them has really worked, and now the tables have turned. Conservationists and scientists have long realised that restoring the bogs and largely leaving them to their own devices can provide more benefits for both people and the environment than any 'improving' management could ever hope to achieve. Now it seems, the world is beginning to catch up.

Sphagnum takes a very long time to decompose. As it dies, and is replaced by new plants, it creates layers of de-oxygenated, partially decomposed soil. This acidic, carbon-rich soil is peat. Peatlands, mostly created by sphagnum, cover roughly 20 per cent of Scotland's land area, some 7,000 square miles. Fifteen per cent of all the world's blanket bog, a particularly interesting type of peatland, is in Scotland.[4] Sphagnum can hold as much as twenty-five times its own mass in water. It therefore increases the water-holding ability of the ground, thus reducing flooding downstream and also keeping off the worst effects of drought.[5] Its water-retaining abilities allow it to alter the ground conditions around it. But it is the ability of peatlands also to sequester huge amounts of carbon that have led to their renaissance in recent years.

Bogs and woods sequester carbon in complementary ways. All those pine trees growing up in the Derry and Quoich accumulate carbon very quickly, between five and twenty times faster than the peat bogs of Bynack, Geldie and Dalvorar. But trees don't grow forever, so the total potential amount of carbon that a woodland can store is much lower than that of a peat bog, although the peat bog will be sequestering carbon at a much slower rate. A healthy peat bog has the capacity to store much more carbon dioxide than a similar area of woodland because the organic matter in peat accumulates over thousands of years. A quick rule of thumb: peat 'grows' at a rate of about a millimetre a year. Therefore those pine stumps buried in two metres of peat are roughly 2,000 years old.

In 2013 the Trust assessed the total peat coverage on most of its upland properties, using satellite imagery and ground surveys.[6] The total store of carbon on Trust upland properties was estimated at 27.5 million tonnes of carbon dioxide equivalent, or the same as about a third of the annual emissions of Scotland. The equivalent figure for Trust-owned woodland was 2.4 million tonnes. Of this, the Mar Lodge bogs were found to hold about a third of all the Trust's peatland carbon stocks: 9 million tonnes of carbon dioxide equivalent. Mar Lodge Estate therefore holds about the equivalent of one ninth of Scotland's annual carbon emissions in its peat bogs. By comparison, its woodlands hold a 'mere' 666,000 tonnes of carbon dioxide equivalent – though this number is increasing year on year.

When it comes to sequestering carbon, it is trees that make the largest short-term impact: woodlands across the Trust's properties capture over 40,000 tonnes of carbon dioxide equivalent every year, while its bogs capture around 15,000 tonnes. In the short term, at least, the importance of bog conservation lies in keeping the carbon that is already locked up in the ground firmly in place. All of this makes the current state of Scotland's peatlands all the more tragic. Because peat is so slow-growing it is very susceptible to erosion. If the sphagnum layer on top is severely damaged by too much grazing or trampling, or by fire, the peat soils may be exposed to the wind and rain, leaching out water and carbon dioxide. Those centuries of attempts to improve the economic viability of our peaty wastelands, often paid for by the taxpayer, succeeded only in wrecking them. Estimates suggest as much as 80 per cent of the UK's peatland landscape has been damaged.[7] Pristine peat bogs will capture around 0.9 tonnes of carbon dioxide equivalent per hectare per year, but a hectare of damaged peat will emit between one and 23.8 tonnes of carbon dioxide equivalent, depending on how much it is eroding.[8] Incredibly, 5 per cent of the UK's carbon emissions are due to degraded peatlands.[9]

There are problems beyond carbon emissions. Peat erosion is bad for water quality. Those tea-stained Scottish burns needn't be as tea-stained as they are. Peat erosion also increases your water bill: organisations like Scottish Water have to filter the peat out of water which accumulates in reservoirs.[10] Peat erosion is also a concern for fisheries: water leaving peat areas is acidified, which impacts on fish survival.[11] Degraded peatlands hold less water, leading to greater fluctuations in water levels in our rivers and burns, and even downstream flooding.[12] The more peat stays in the uplands, the better.

Historically, it was state-sponsored drainage of our peatlands and their afforestation with non-native forestry that caused the greatest problems for our sphagnum-rich bogs. The activity which does the most damage to our peatlands in the here and now is the practice of muirburn on peat soils. The burning of heather promotes the growth of young shoots at the expense of rank, old heather. This is done to create a food source for grouse, which will subsequently be shot for sport. Muirburn creates the open, heathery moorland landscape we all know from the postcards, but it can be highly damaging to peat and emits a huge amount of carbon in the process.[13][14] Muirburn was traditionally undertaken only on mineral soils, but in recent decades, as grouse moor management has intensified, the practice has encroached onto peat soils.[15] In spite of all this, a recent study seemed to show the muirburn doesn't actually increase grouse numbers all that much.[16]

'The muirburn debate' sometimes feels like a new development in the wider debate about our moors. After all, people have been setting fire to the hills for centuries. It has a long history at Mar Lodge, and we know about this partly because of all the fuss that it causes. In 1725, the new (and somewhat temporary) owners Lord Grange and Lord Dun brought in regulations to limit its use. They were worried about the potential impact of burning on the valuable pinewoods. Sixty years later, a court stipulated that muirburn could only take place at Mar Lodge between November and April – a more stringent law than is in place today.[17]

One argument which continues to do the rounds in defence of muirburn on peatlands is worth a closer look. The argument goes that managed 'cool burns' created by muirburn reduce the risk of catastrophic, extremely hot wildfires starting by reducing the fuel load of heather, leaving underlying sphagnum mosses intact.[18] Further, muirburn creates firebreaks, which stop the spread of catastrophic, large wildfires.[19] It's a neat idea, and in some circumstances reducing the risk of wildfires might be an acceptable price to pay for the detrimental impacts muirburn can bring. But regular muirburn creates conditions in which only fire-loving species like heather can thrive, to the detriment of all the other species. If you repeatedly burn and drain a peatland, soon all you are left with is extremely combustible dry peat, and species which don't mind being burned, like heather. This means that the more you burn an area, the more combustible it becomes. To suggest that muirburn improves the lot of sphagnum mosses, plants which need lots of water to survive, is ludicrous.[20] Even worse, internal National Trust for Scotland research, quoting figures from the Game and Wildlife Conservation Trust, suggests that as many as half of the wildfires in Scotland are ignited through muirburn running out of control.*[21]

Recent events south of the border have put the effects of muirburn on peatland, particularly wet bogs, further in the spotlight. In 2004, RSPB took over management of Dovestone, near Saddleworth Moor. The bogs were degraded, having been dried out by decades of

---

* As far as internal reports go, this one is remarkably direct. It states that 'it is therefore perverse to recommend [muirburn] as a means of reducing the risk [of wildfire]'.

draining and burning. Instead of continuing with muirburn, RSPB staff diligently spent a decade filling drains, reseeding sphagnum onto exposed peat areas and bringing the water levels back up. As work progressed, more waders started breeding, drawn in by an abundance of craneflies, which were thriving in the newly wet bogs.[22] The dunlin population went from seven pairs in 2004 to forty-nine pairs in 2017. Curlew numbers went up from twenty-seven pairs in 2010 to thirty-six pairs in 2017. In total, breeding waders went from ninety-three pairs in 2004 to 195 pairs in 2017.

Then, in 2018, following extremely dry conditions, the adjacent grouse moor, Saddleworth Moor, went up in flames, burying Manchester in smoke for several days.[23] Despite the regular muirburn prescribed by moorland managers, almost the entirety of Saddleworth Moor was burnt. The fire drove along the uplands until it got to Dovestone, where it was stopped by wet blanket bog. The only area of RSPB land that did burn was the degraded peatlands, the stuff that was still in a state of damage from historic burning and draining.[24]

The peat bogs of Mar Lodge have not been burned for many years, but in 2017, the Trust committed to stopping muirburn entirely at Mar Lodge.[25] It was a remarkably courageous act for a historically somewhat quiescent organisation; muirburn is ingrained in the minds of many moorland managers and owners as an essential part of maintaining a healthy moorland. It is as much a cultural, performative gesture as it is an act of land management, and by ending the practice the Trust was risking its reputation with other sporting estates. Ironically, that courage may have been unnecessary. In October 2019, then Tory MP and now member of the Lords Zac Goldsmith told MPs that the government agreed that burning peatlands was a problem: 'There has been an attempt, through voluntary initiatives, to scale back – to reduce and eventually eliminate – the burning of fragile and important peat ecosystems, but that has not proven 100 per cent successful as had been hoped. We are developing a legislative response to the problem.'[26] In January

2020, the Committee on Climate Change, the UK government's independent advisory body on climate change, recommended that the UK should ban muirburn on peat soils, while also restoring 50 per cent of degraded upland peatlands.[27] According to their figures, 'restoring at least 50 per cent of upland peat and 25 per cent of lowland peat would reduce peatland emissions by 5 megatonnes of carbon dioxide equivalent by 2050'. This action would be the carbon equivalent of taking around 1 million cars off the road.

Other sphagnum protection work is rumbling along in the background at Mar Lodge. Previous work has been done to block up ditches which were dug in the 1980s, as part of a scheme to 'improve' the land. Reducing deer numbers has had its own knock-on impact, reducing erosion caused by trampling. The next phase is restoring the most degraded, 'haggy' areas of Mar Lodge's peatlands. This will involve damming and reprofiling hags, some of them several metres deep, with specialist kit, and then reseeding the area with sphagnum and heather. This is part of the Cairngorms National Park Authority and NatureScot's Peatland Action Program, which is funded by you, the taxpayer.[28]

All of this is good news for dunlin, and for you and me too. More peatlands in better conditions lead back to wetter bogs which are more resilient to climate change. These in turn protect us from climate change and its effect, by sequestering more carbon and reducing flood risks. All the while, wet bogs mean more dunlin, whose larger populations are more resilient to climate change.

The sheer inhospitality of Scotland's vast peat bogs challenges the comfortable idea that the world exists to furnish humans with stuff. They challenge even the idea that our wilds are places which exist so that we can have great adventures in them. Exploring bogs is slow, wet, uneventful, tedious, and rarely photographs well. The bogs have defied our best efforts to tame them for centuries. Now, belatedly, we have realised that they have always provided for humans, by filtering our water, providing homes for the wildlife we share our land with, soaking up our carbon. They are humbling landscapes, which know our needs better than we do ourselves.

It takes six whole glances to see the real face of Carn an Fhidhleir, and fall in love with its boggy slopes. This sixth glance is the humble look that filters out human prejudices about what wild places should be like, and what we can get from them. It sees the place as it really is. It sees the tiny dunlin, sitting up on a hummock of sphagnum, proudly sporting black, white and polished chestnut, singing loudly for all to hear, 'I'm a dunlin, this is my home.'

9

Atlantic Salmon

> They say that he was led to invade Britain by the hope of getting pearls, and that in comparing their size he sometimes weighed them with his own hand.
>
> Suetonius, *The Life of Julius Caesar*, AD 121

One of the joys of environmental conservation is the rule of unintended consequences, the cascading butterfly effect that environmental work can bring to not just one species but whole ecosystems. And few species link places and people together more than the Atlantic salmon.

May 2019, during the only blisteringly hot weather of the year, and a group of volunteers heads out to the Geldie. It is a largely treeless glen, bar an ugly square Scots pine plantation, planted before Trust ownership, which sits slap bang in the middle of it. The glen straddles the regeneration zone and the moorland zone, acting as a buffer for woodland regeneration efforts in Glen Dee. Historically, it has been home to hundreds, sometimes many hundreds, of deer. Now, there are fewer.

I said it was largely treeless, but that's not entirely true. One 'tree', dwarf birch, covers a huge swathe of the northern flank of the glen, and a fair chunk of the south-eastern edge as well. It is a remarkably pretty plant, with tiny, almost semi-circular dark leaves and golden ginger catkins. The dwarf birch in the Geldie is, we think, part of the largest dwarf birch forest in Britain. But this forest is all less than knee-high, sitting below the heather: dwarf birch is highly palatable, and is happily munched away by deer.

A word about dwarf birch. In 2016 I went hiking in northern Sweden, through the giant patchwork quilt of pinewood, birchwood, montane heaths, mountain passes, lakes, rivers and glaciers that comprise the area around Abisko. Here, the woods were of the montane kind, which we'll be coming back to in a later chapter. Mosquito-infested downy birch joined willows along the rivers and lower glens, protected from reindeer and elk by late-lying snow. Patches of twinflower and small cow-wheat thronged the forest floor, while bluethroats, redwings and bramblings sang us along our way. Below the downy birch sat thickets of dwarf birch, with its tiny rounded leaves and dark twiggy branches.

Walking through this remarkable landscape, it took me a while to realise that this dwarf birch was the same as the stuff I had come across in Scotland. In Sweden it formed large, often impenetrable bushes, growing everywhere, up to a couple of metres high, acting much like heather does in Scotland. It was dripping in birds: reed buntings, Lapland buntings, ring ouzels, blue-headed wagtails. Willow grouse, the same species as the red grouse we have in Scotland, used it as cover and ate its buds. Whenever I had come across dwarf birch in Scotland, I had stumbled on it in bogs, sitting low, hiding out below the heather, a stub of a tree, so small that not even a meadow pipit could perch on it. It was a revelation. It got me to thinking – what would this place look like if the dwarf birch returned to the Geldie?

The volunteers are restocking a new patch of woodland, creating a broadleaf link between the sorry-looking square of pines and the Geldie Burn. Here, within fencing (too many deer to hope that trees could establish without one), are some 30,000 trees, planted in 2017. The soil here is poor, leached and acidic, suffering from centuries of high levels of browsing by deer. It is exposed. The trees were planted without the use of fertiliser. Mountain hares have taken their toll, killing some of the trees, and so the planting area is to be restocked with a mixture of broadleaf species, aspen, downy birch, alder, rowan and four species of willow, to make up for the mortalities.

My goodness, it is hard work. There are twelve of us. The volunteers are a hardy lot borrowed from the John Muir Trust, and we are aiming to plant around 3,000 trees. There are no bridges, so the Geldie Burn must be forded and reforded. The sun beats down on burning necks. The ground doesn't dig easily, it must be screefed. Sandwiches are wolfed down, followed by regret that there are no more sandwiches to be had. But spirits remain high, even as the planting rate seeps downwards as the day goes on. People love planting trees.

All of this activity, one way or another, is for the benefit of salmon. We may think of salmon as a game species of the open, treeless moor, but people are becoming increasingly aware that salmon need trees to survive. Trees regulate water levels and temperatures, they stabilise riverbanks creating gravelly patches.[1] They flood the waters with nutrients, which leads to more insects.[2] They are also a major ally in downstream flood prevention.[3] Above all other species, salmon show us that we should think of the moors as a patchwork of open ground and wooded spaces. The salmon would find the arbitrary constructs of the tree'd woodland and the treeless moor to be simply incomprehensible.

Salmon may benefit from the trees we are planting, and the trees themselves do pretty well out of it. But the actual, final intended beneficiary of this activity, the species which is actually paying for it all, is another beast altogether. It is a curious creature that people seem to have largely forgotten about, and it is dependent on salmon for its own survival.

The freshwater pearl mussel is one of those bizarre creatures that simply defy comprehension. To look at, they are like the mussels you get in Belgium when you order moules frites, except they're a bit larger and not covered in garlic and butter. They can live for over a century. In fact, they have a maximum lifespan of around 250 years. They spend their lives sitting on gravel banks, usually a metre or so below the water, filtering the river, cleaning it, and living off the gunk that they filter out. Also, one in every few thousand has a pearl (yes, a real pearl!) in it.

It is odd indeed to think that Rome's first emperor decided to come, see and conquer Britain for the sake of pearls. It is, if nothing else, a stark reminder of how much we have lost. Even more remarkably, it was only fairly recently that many people were aware of them. The last full-time pearl fisher, a chap called Bill Abernathy, was working the Esk until the 1970s, floating about in a coracle made of canvas and sticks, looking for them through a special, giant snorkel mask. His most famous find, the marble-sized Abernathy pearl, is the 'jewel in the crown' of Scottish pearls. Its current owners, Cairncross of Perth, rather coyly describe it as being 'priceless'.

The reason that you can't buy Scottish pearls so often these days, and certainly can't (legally) fish for pearl mussels, is that we killed pretty much all of them. And yet, despite this, Scotland has some of the healthiest freshwater pearl mussel populations in the world. More than half the world's recruiting population (still reproducing) exists in Scotland.[4] We killed most of them, but we weren't quite as good at killing them as other places were at killing theirs.

For most Scottish people, the freshwater pearl mussel is an unknown, forgotten relic of history. For the few who do know about it, all too often it is the lure of those pearls. A pearl poacher can kill an entire population in a single search for riches.

Pearl mussels may be largely forgotten, but they perform important ecological functions that benefit us all. They filter water of pollutants. Mussel beds can strengthen riverbanks, stopping erosion. They also provide food for any number of freshwater species. But one particularly potent reason that we are interested in pearl mussels is that they spend the first nine months of their lives latched on to the gills of salmon and trout. Without salmon, they cannot survive. So any conservation project to save them will have to save the salmon first. And people like salmon. Salmon are beautiful. Salmon are a keystone species of Britain. They are delicious. Salmon fishing is highly lucrative. We have a cultural connection with salmon that goes back to the beginnings of civilisation, one that, unlike the freshwater pearl mussel, has yet to be forgotten. The salmon is a conduit: protect the salmon and you protect the pearl mussels, and vice versa.

All of this is by way of saying that freshwater pearl mussels were paying for our tree-planting efforts. The Geldie planting scheme is a small part of the hugely ambitious Pearls in Peril project, a multi-million pound, EU-funded project to protect pearl mussels for the benefit of pretty much every species that they share the rivers with, not least salmon.

That's not where the potency of the salmon brand ends. As we are slaving away under the hot sun, I become increasingly aware that there are not just planted trees here. Despite the apparent lack of a seed source, there is natural regeneration here too, and it is doing much better than the planted stuff. A few willows and rowans, unlocked from the perpetual cycle of growing above the heather and being browsed back again, are roaring away. I keep finding dwarf birch plants. They must have tripled in height since the fencing went up two years previously. They are rocketing above the heather. Insects flood onto the spring flowers of eared and creeping willows, each catkin a riot of life. The volunteers dunk their heads in the burn, reapply sun cream, and get back to planting. Two thousand trees to go. My mind goes back to Sweden, the dwarf birch forests, and I wonder what Glen Geldie might look like in fifty years' time, and what creatures might be at home in it.

And here is the beauty of a simple conservation intervention. Volunteers plant trees, which stabilise riverbanks, reduce river temperatures, sequester carbon and return nutrients to the river, helping more salmon to survive. More salmon means more homes for pearl mussel nymphs. Pearl mussels, in turn, filter the water, further stabilise banks and provide food for any number of species. Finally, the trees are then helped by returning salmon, who bring back nutrients from the sea and deposit them on land, by way of ospreys and other predators. Planting trees helps salmon, salmon help freshwater pearl mussels, which help other river species, which help salmon, which

Mar Lodge Estate: 30,000 hectares of regenerating Caledonian woodlands, rolling moors, bogs and subarctic mountains (Andrew Painting)

Mar Lodge. Queen Victoria laid its first stone in 1895 (Andrew Painting)

Above. The pinewoods are returning to their former glory in the Derry, one of Scotland's finest glens, thanks to a reduction in deer numbers (Andrew Painting)

Left. The Linn of Dee, a popular beauty spot where the salmon still run (dvlcom, Shutterstock)

The second largest known Scots pine in Scotland, surrounded by regenerating pines, birches and rowans (Andrew Painting)

The oldest known pine at Mar Lodge, dating back to at least 1477. It has seen the loss of the pinewoods from this glen, and now is seeing their return (Shaila Rao)

Young pines and birches are crowding along the length of Glen Quoich for the first time in centuries. Behind looms Lochnagar, ten miles distant (Andrew Painting)

The regenerating Caledonian pinewoods of Mar Lodge, where the area of native semi-natural woodland has more than doubled since 1995 (Andrew Painting)

Buxbaumia viridis, the green shield-moss. The giant panda of the moss world? (Shaila Rao)

A black grouse lek, early in the morning, where males duke it out to impress the greyhens (Shutterstock)

Left. The Kentish glory, a moth designed by an abstract expressionist, is making something of a comeback in the Cairngorms National Park (Patrick Cook)

Below. One of the lynx at the Highland Wildlife Park, Kincraig. This is the only place in the Cairngorms, and indeed Scotland, where you can see this native species (Highland Wildlife Park)

Shaila Rao, ecologist at Mar Lodge since 2002 and a driving force behind the restoration of the Caledonian pinewoods (Andrew Painting)

'The corpses of thousands of midges mixed with the rain, creating a soapy paste that threatened to leach into the interior workings of the computer.' Ranger Ben Dolphin on a typical August day (Andrew Painting)

A stalker takes a stag off the hill with a Highland pony (Chris Murphy)

The gamekeepers clear up after a wildfire which ripped through Glen Luibeg in 2014. The cause was found to be an errant campfire (Chris Murphy)

The ruins of Geldie Lodge sit below An Sgarsoch, in a vast area of moor and bog (Andrew Painting)

Red deer and woodlands belong together, but thanks to humans, they now have a somewhat dysfunctional relationship (Chris Murphy)

A satellite-tagged hen harrier. The species has made a dramatic return to Mar Lodge but suffers from illegal persecution elsewhere (Shaila Rao)

Left. A dunlin nest – blink and you'll miss it (Rab Rae)

Below. Sphagnum moss, a keystone of bog habitats and one of our best allies in the fight against climate change (Andrew Painting)

Right. Salmon make their way up the Linn of Dee, as they have done for as long as the Linn has existed (Philip Lay)

Below. The estate maintains a herd of around 1,500 red deer at Mar Lodge, and Highland sport is an important part of the estate's management and ethos (Chris Murphy)

The stag ballroom, housing 2,435 heads, is a temple to the cult of the red deer (Andrew Painting)

Winter in the subarctic Cairngorms, the highest, snowiest mountains in the UK (Andrew Painting)

Above. The Sphinx snow patch, the longest lasting snow in Scotland, here at its annual low ebb in September 2020 (Andrew Painting)

Right. Alpine sow-thistle, one of Scotland's rarest plants, has been restored to Mar Lodge by a team led by the Royal Botanic Garden Edinburgh. This is the first flower recorded on the estate in the 21st century (Andrew Painting)

Hutchison's bothy, Coire Etchachan: 'So will I build my altar in the fields,/And the blue sky my fretted dome shall be' (Shaila Rao)

Red House bothy, with historic muirburn behind. The Trust discontinued this controversial management technique in 2017 (Andrew Painting)

The author (right-hand dot, near centre) on Sputain Dearg (Garry Cormack/hillgoers.com)

The dotterel, imperilled by climate change: 'these lovely and confiding dotterels will always call and hold us. We remember them with affection' (Shaila Rao)

A golden eagle chick on the nest in a remote patch of Mar Lodge's Caledonian pinewood (Ewan Weston)

A worker monitors a precious scrap of downy willow, part of a lost ecosystem making a tentative return to the Cairngorms (Adele Beck)

fertilise trees. Meanwhile, the planting of trees has the additional benefit of creating new environmental niches for any number of terrestrial species, from black grouse to mountain hares. And to top it all off, dwarf birch, one of the UK's rarer plant species, gets its own kick-start. We humans not only get to enjoy living with a more diverse and intact ecosystem but also benefit from flood prevention, less water pollution, carbon sequestration and the economic bonus of more salmon to fish for. Meanwhile, landscape-scale environmental restoration affects the way that all of us, from land managers to visitors to people who flush their toilets, think about our river systems, what they look like, what they could look like in the future, and how we should care for them. And the salmon, the king of the river, is at the heart of it all.

Of course, this is all pie-in-the-sky thinking. Far too neat and ambitious. The Pearls in Peril project has done an enormous amount of good for Deeside and beyond. But on its own, it will not save either the Atlantic salmon or the freshwater pearl mussel from what are amounting to catastrophic declines across the whole of Scotland.

I'm not an angler, at least not one of any great talent, so the salmon I see are glimpsed, a quick flash up the Linn of Dee, a jump and splash in the river. Very occasionally I come across salmon carcasses, gnawed on by opportunistic otters. But even these half encounters are laced with a specialness. Salmon truly are the king of fish. They have quintessence.

Your average punter will be surprised how big salmon can grow. The largest fish caught in the Dee was a whopper: fifty-seven pounds, and over four feet long (fish are always measured in pounds).[5] The biggest caught in Scotland was on the Tay, a sixty-four pounder caught by Georgina Ballantine nearly a century ago.[6] The cast of a forty-eight pounder sits proudly in Mar Lodge. In 2013, a fifty pounder was caught on the Tweed, another famous salmon river. They are impressive stories, yet they are also laced with a nostalgia for a better time in angling, when there were more, bigger fish.

It's July, and Pamela, Bas and Eilidh are visiting from the River Dee Trust. This organisation is the one that looks after the Dee's salmon and other fish, and this year it has been tasked with a fairly rigorous schedule of fish monitoring. This afternoon's plot is on the Lui. It happens to be a nice spot, a big meander with a broad sandy beach surrounded by regenerating pines. It's a popular picnic spot with day trippers and campers. It's not raining for a change. The sky is moody, but there is a breeze lifting both the midges and the close muggy heat that has clung to this part of the world for the last week. It is, all in all, a pleasant day for a paddle, and that's exactly what Pamela, Bas and Eilidh are up to.

Bas, a tall Dutch statistician, is attached to a Ghostbusters-style piece of kit – a large rucksack attached to an anode on a hand-held stick and a cathode trailing in the water. Pamela and Eilidh are slightly downstream. They are electrofishing. Bas sends a small current through the water, enough to stun fish, which are then collected by Pamela and Eilidh. It's a slightly frenetic affair; lots of kit is wielded awkwardly in waders, and the fast-flowing water brings an extra element of jeopardy to think about. They zigzag upstream, methodically, making sure that each bit is given equal attention.

I'm watching from the sandy bank, stymied from helping out by a lack of waders (and, perhaps, skill; like Rob Wilson's tree-coring team, this is clearly a competent group of people and the last thing they need is incompetent assistance). A common hawker dragonfly, one of the first I've seen this year, hawks by. An equally busy but somewhat more floaty small pearl-bordered fritillary searches for nectar. I'm reminded of Jerome K. Jerome, 'I like work: it fascinates me. I can sit and look at it for hours.'[7]

Salmon and sea trout hatch from their eggs, which have been laid in redds – gravelly nests in the riverbed. First they are fry, then they grow and become parr, about a finger long. Fish are slow-growing in the Upper Dee catchment. Up here it takes a couple of years, maybe three, for the fish to become smolts, the point at which they decide to chance their arm out at sea. They migrate to the open

Atlantic, undergoing remarkable physiological changes to cope with the salinity of the water. Here they wander about for another couple of years before deciding that it is high time that they settle down and have a family. So they return to the river that they were born in, somehow, and indulge in a fish orgy, and then they die. The chances of any individual salmon fry completing this full life cycle are tiny. It all sounds exhausting. I could sit and look at it for hours.

By and by, the team's allotted hundred square metres are surveyed, and the return is slightly, though not wholly, disappointing. Here are four brown trout parr. They are beautiful – spotted and richly coloured, quick and powerful, vivacious. 'Probably a lot older than how they look,' Bas explains. 'Up here, it's a harder life.' Brown trout don't bother with going to sea, they just hang about in the river their entire life.

Eilidh measures the fish and carefully pours oxygenated water over them. Pamela writes the particulars, identifying them to species – trout and salmon fry look similar, but there are a few telltale signs; the shape of the tail, the length of the mouth in relation to the eyes. 'Sometimes you can tell by colour, but not always,' says Pamela.

Pamela studied ecology at Aberdeen University and joined the treadmill of seasonal environmental work, picking up a summer contract with the River Dee Trust. The work involved walking the length of the Dee, mapping it in incredible detail, noting spawning areas, barriers to fish migration, points of erosion. It was an extremely useful exercise, and she's now been full-time with the River Dee Trust for a couple of years, working on a whole array of monitoring and conservation measures designed to improve both the fortunes of salmon and our knowledge of what is troubling them. This work in the Lui will be fed into a national dataset.

Surveying in the uplands can be hard-going, Pamela explains. Up here the water is oligotrophic; there is little in it that conducts electricity, so the voltage on Bas's backpack has to be pumped up. You could be generous and say that this means the water is extremely pure, or mean-spirited and say that it is lacking in life. Both would be true, at least to a certain extent.

Seeing young trout is a good sign of a healthy river, but the Lui is not a salmon river. There are powerful waterfalls that act as a barrier to returning salmon. In the 1960s, the estate owners saw an opportunity to increase the amount of fishing they could sell and installed a salmon ladder. This is a series of concrete boxes next to the falls through which water could flow downstream, but up which salmon could, in theory, navigate. Unfortunately, it never really worked, and the work left an ugly concrete scar on the rapids. Eels, equally endangered, can make it up the Lui waterfalls, but not salmon. To find them we need to head out to the moors.

Mar Lodge sits some fifty miles from the sea, almost as far inland as salmon can run in Britain. For this reason, it doesn't have quite the same illustrious cachet (or price tag) as other Deeside fishing beats, but it does offer some of the wildest and most challenging fishing sport that you can get in Scotland. Salmon fishing contributes £15 million a year to the Dee catchment. 'It's one of the well-known rivers. It's a big thing to fish the Dee,' says Pamela. Almost all of our knowledge of the Dee's salmon stocks comes from anglers. Fish stocks wax and wane over successive years, but it's thanks to their figures that we know that rod catches on the Dee have declined from around 14,000 in 1957 to 3,500 in 2019. While there have been some vintage years for angling in the not-too-distant past (2010 saw a catch of around 10,000, for example), for the most part, as has happened across Scotland, salmon numbers have been in freefall.[8] For every 100 salmon that leave our rivers, fewer than five return – a decline of nearly 70 per cent in just twenty-five years.[9] 2018 was the worst year in history for salmon returning to UK waters.

Salmon are indeed facing a lot of troubles in our rivers. A lack of coarse woody debris* in the upper catchment, caused by historic deforestation, has led to a loss of nutrients, destabilised riverbanks

---

*   A jargon term for 'tree branches'.

and increased sediment build-up in this area. The lack of shading caused by this historic deforestation exposes the upper moorland rivers to the sun, and so introducing woody debris will introduce diversity to the river, creating deeper, cooler pools, which are a refuge for fish in warm weather. Further downstream, there are now lots of human-made obstacles for fish to navigate, like weirs and hydroelectric dams. Water pollution from agricultural run-off like cow manure and fertiliser causes eutrophication, or too much nitrogen in the water, while pesticides kill off invertebrates. Abstraction of water for human use lowers river levels in times of drought. In the past, overfishing may have been a problem, but now the Dee is a catch-and-release river. Then there are the anglers' bogeymen, predators like cormorants, goosanders, otters, seals and dolphins. The return of these species after long-term declines has been hailed as a conservation success story. Now, these species are once again being blamed by some anglers for the demise of the salmon.[10]

There is no denying that all of these species are predators of salmon. But it cannot be the case that in a fully functioning ecosystem predators will eat themselves out of a food source. The control of native predator species can never be a silver bullet for environmental conservation, but it can provide a temporary respite for a particularly beleaguered species. While predators like goosanders have a safe home at Mar Lodge, further downstream goosanders are legally shot under licence. The efficacy of this measure is hotly contested, the practice highly controversial.

The River Dee Trust is a regular fixture at Mar Lodge. It looks after temperature loggers installed in the Geldie Burn, which are part of a national array set up by Marine Scotland Science.[11] The River Dee Trust also conducts annual salmon redd (nest) counts, alongside their electrofishing endeavours. This leads us to yet another problem: climate change. Water temperature was a huge problem in 2018 and may well be again in the future. The Dee needs snow on the hills to keep the water topped up and cool through the summer. In 2018 the River Dee Trust had to delay some of its survey work, as the water temperature was too high for the kit to work effectively. The temperature of the water got to 26°C in the Geldie; 27°C is lethal to salmon. This might become the new normal. This is another reason why the Pearls in Peril tree-planting programme is so important. The trees will provide shade, cooling the water. 'One-off' events like Storm Frank, which ripped through Deeside in 2016, have had huge, but mostly short-term, impacts on fish stocks. However, with climate change, we can expect these events to become more frequent and more damaging.

The problems don't end there, however. Salmonid problems are not confined just to the rivers. Overfishing of wild salmon and their prey at sea brings more pressure. Meanwhile, particularly on the west coast, a dramatic rise in farmed salmon is causing outbreaks of sea-lice infestations in wild salmon.[12] If the great joy of conservation is seeing the pieces link together, then its opposite is seeing how complicated can become the task of saving a species which faces multiple, often contradictory, threats. For our rivers, it's a vicious cycle – degraded ecosystems will lead to reduced fish stocks, while reduced fish stocks lead to further degraded ecosystems. The trick is to act before it is too late.

It's October, and the salmon are leaping, moving nutrients and calories between the high seas and the high hills. I stand on the bridge over the Linn of Dee, looking down into the narrow gorge, not much more than six feet wide in places, through which all of the Dee must transport itself. The water level can change here by a matter of dozens of metres between high and low water. It is a formidable barrier, but the salmon manage it just fine.

It's often busy here. It's a popular beauty spot, suitable for wild camping, with several swimming holes a few hundred metres downstream, but today I have it to myself. The birches are gold, the larches are turning, the blaeberry is fire-red. I blink, and a salmon jumps upstream. It seems like a small miracle, but it has been happening here for the whole lifetime of the Linn of Dee. It is as normal an event as water flowing from the mountains into the sea.

It's certainly not too late for salmon. In the office we have pictures from 2013 taken by intrepid scuba divers of salmon sitting about below the Linn of Dee, the water crystal-clear, mustering their forces for an assault on the Linn. Chris, the head stalker, tells me about the 2010 season, when salmon returned in their droves, too many to count, hundreds thronging even the smallest burns, a cacophony of life.

The fight is on. The Geldie planting site is just a tiny part of a huge suite of actions going on across the entire Dee catchment, helping salmon, pearl mussels, even dwarf birch. The River Dee Trust has already planted 200,000 native trees across the moors of Deeside, but it has now committed to planting a further 800,000 by 2035.[13] Elsewhere on the Dee, the River Dee Trust is adding nutrients to burns by simply chucking deer legs into them. An army of volunteers are removing non-native invasive species from the downstream watercourses, including Himalayan balsam and giant hogweed. Mink rafts, clay pads which record the footprints of non-native mink, are installed and monitored by another army of volunteers. Water bailiffs scour the riverbanks for poachers.[14]

Down on the Tay, beavers have returned for the first time in centuries. Their dams are bunging up the rivers, slowing the water flow, creating new places for fish fry to thrive, sifting out pollution, encouraging a natural flow of water. The beaver is the salmon's ally, and indeed the ally of all freshwater creatures.[15] For now, for the Dee's salmon, beavers might as well be a thousand miles away. It will take a long time for them to naturally spread from the Tay to the Dee, and there is a moratorium on beaver reintroductions, brought about in part by pressure from farmers and, believe it or not, anglers, caught up in a myth that beavers are bad for salmon.

Growing trees is at the low-tech end of a suite of salmon conservation measures going on across Scotland. Meanwhile, at the high-tech end of the conservation spectrum, an array of salmon-spotting beacons have been installed around Aberdeen harbour. Tagged salmon are picked up by the array, telling scientists how many salmon are reaching the sea and where they are doing so. An even larger project using similar tech is underway around the Moray Firth, involving seven river catchments, including the Spey. This project will synthesise what is known about potential losses and create a complete management framework for salmon across a huge area of Scotland.[16] It's an expensive project, costing millions every year, but with the economic benefits of salmon fishing at stake (fishing the most prestigious stretches of rivers like the Dee or the Tay for a week will set you back a five-figure sum), there are plenty of people out there willing to stump up the cash to support it.

Another salmon catches my eye, flashing quicksilver, jumping up the river. Its whole life is one long exercise in forlorn hope.

The reason salmon are so important is their mobility – it is their remarkable life cycle that engages them with so many other different creatures and habitats. But this is also their weakness. For salmon to survive, a whole chain of ecosystems need to be linked, and to remain in good functioning condition. For that to happen, things must come together across not just an entire continent but an entire ocean as well. It is an uncomfortable reminder of the limitations of nature reserves. However big they may be, however ambitious their schemes, nature reserves alone will not save our imperilled environment. Mar Lodge's moors are just a small part of a salmon's continent-wide home. To save them, we must learn to think more like salmon, allowing cultural landscapes and natural ecosystems to merge into one another. We, too, must learn to merge ourselves back into their rhythms and needs.

10

Curlew

'The best part of ornithology is nest finding,' says David Jarrett of the British Trust for Ornithology (BTO), as we scan the expansive Quoich Flats for lapwings on a cold, sullen April day. 'It's a skill that's in danger of falling out of fashion, what with all the complicated modelling we can do these days. You have to look to the older papers and books to find the best nest-finding tips.'

The Quoich Flats, which form the floodplain that links the Quoich Water to the Dee, are particularly beloved by waders. Lapwings, piebald, quiffed, with stubby bills, iridescent backs and peculiarly shaped wings, take shelter in the rough pasture on the higher ground. Snipe (small, mottled brown, with an extraordinarily long bill) and redshank (slightly larger with bright red legs and bills) flock to the wetter scrapes. Oystercatchers and common gulls use the shingle beds, deposited by the storms of 2014 and 2016. Displaying lapwings throw themselves into the air, singing their weird, radio-tuning song, 'peewit, peewit'. Over in the far distance, the best part of a mile away, a curlew sings, 'whaup-whaup-whaaaa-uuup curloo curloo'.

Few birds evoke a landscape more thoroughly than curlews evoke Scotland's uplands. Large, long-legged, with enormous curved bills, curlews live and breed in the moors and sheep walks. Their bubbling song has lifted the spirits of successive generations of keepers, farmers and walkers. By turns gregarious and obscenely secretive, subject to curious fits of shyness, they are somehow both wild and friendly. For many in the uplands, having breeding curlews on your land is a marker of success – it shows that you are sitting lightly on the land, while also bringing out the best that it can offer. The return of curlews to the uplands in spring is one of the highlights of the season. Of all the waders that frequent our uplands, none has been taken more to heart than the curlew.

Nest finding is hard, but it is also rewarding as only difficult, skilful work can be. It is the sort of activity that crosswords and sudokus aspire to: a primordial puzzle, the sort of thing that you feel humans are designed to spend our days doing. Finding bird nests requires a deeper knowledge of your quarry than the average birder possesses. David points out nuances of bird calls, timing, habitat requirements, how much cover to sit in, how close to get. Nest finding requires a deeper attention than most of us get the chance to exercise in our daily lives.

Then there's the eggs themselves, intricate incarnations of nature, hot to the touch, infinitely varied colours and patterns. Egg collecting was banned in 1954, and rightly so, after it pushed several species to extinction in the UK. But egg collecting is only ever acquisitive – it reduces an elemental activity to the level of stamp collecting. Nest finding for scientific reasons, provided that the data recovered will be more beneficial to the species than the minimal stress caused by an intrusion to a nest site is detrimental, is one of the great joys of spring.

David spends most of his spring nest finding. In 2019 one of his big jobs is working with keepers and rangers across the eastern Cairngorms to monitor wader numbers. He's talking to three of us: myself, Ben Dolphin (seasonal ranger, journalist and president of the Scottish Ramblers – most conservationists have fingers in several pies, simply because there is so little work to go round and the pay is so poor) and Jos Milner, the East Cairngorms Moorland Partnership (ECMP) officer. The Partnership, a collaboration between the Cairngorms National Park and six sporting estates, promotes best practice in moorland management and facilitates the delivery of public benefits alongside private interests. The estates, of which Mar Lodge is one, have targets which include expanding woodland and scrub cover, improving the conservation status of birds of prey, and restoring blanket bog habitats. The estates encompass a truly enormous area, around 140,000 hectares. Jos has been running the wader project on the ground, getting staff up and down the eastern Cairngorms out finding wader nests. 'It's not hard,' she tells me. 'Keepers love curlews.'

David shows us how it's done with a lapwing nest. We peer through a telescope that he's set up at the edge of the Flats. It is focused on a lapwing. She's looking shifty, her head just about popping up through the long grass. She's clearly sitting on a nest, but actually locating it in the field is difficult, even when you can see her sitting on it. David pinpoints it in the landscape, triangulating her position using markers, a molehill, a tuft of grass, a telegraph pole, and we walk up to where he thinks it should be. The lapwing susses what's going on, flies off with an indignant 'peewit', sending other lapwings into the air as well. Even still, the nest remains elusive for a couple of minutes. We tread carefully around the ten-metre circle we think it's in, before David offers a contented 'ah!' and points to the ground. There, in a tussock of grass, is a beautifully weaved cup containing four speckled eggs. We take a grid reference and a photo and David deftly places a small black lozenge below the eggs.

'It's a temperature logger,' he explains. 'We want to know how many of our waders are hatching, and how many are being eaten, and what is eating them. Temperature is a great marker of that. If the nest hatches successfully, the temperature declines inconsistently over about a 24-hour period, as the chicks hatch and gradually leave the nest site. But if a predator eats the eggs, the temperature in the nest will drop suddenly. If it happens at night, it's highly likely that a mammal will be the culprit, a stoat or a fox. If it's during the day, it's more likely to be a corvid or a gull.' The lapwings swirl around, peewitting away, and we retreat back to the telescope. The distant curlew, unfazed, continues its singing.

David clearly loves waders, but he maintains a rigid scientific impartiality towards their plight. Total impartiality is something that the BTO prides itself on. He entrusts myself and Ben with a handful of temperature loggers and sets us nest finding for the remainder of the season. It takes us a while to get the knack. Between us, we realise that it is best to work in tandem. One of us uses the telescope to find a nest, and keeps it focused on the location. The other person walks towards where they think it will be. They are then guided to the nest through the use of elaborate hand gestures from the telescope operator, a sort of semaphore-style system. Sometimes, it even works.

Ben and I keep a close eye on the waders over the spring and summer. We do wader transects, in conjunction with the nest finding. This is just walking about, looking (and more importantly listening) for waders, noting their behaviour and age and jotting them all down on a map. The BTO is keen to try out more than one survey technique at a time – if it turns out that an easier way of monitoring wader breeding success produces the same results as more labour-intensive projects, then so much the better.

We're certainly not the only ones working on the project. Across the ECMP area, in 2018 and 2019, an impressive 183 nests are found by gamekeepers, rangers and ecologists. Alongside the temperature loggers, we also use trail cameras to keep an even closer eye on a select few nests. As well as lapwings, we find oystercatcher nests, delicate gravel constructions, lovingly arranged grains of grit and pebble. They are easy enough to find – oystercatchers are not the most subtle of creatures. One oystercatcher makes her nest right next to a footpath. She's there every year, extremely conspicuous in black, white and red. Her nest is a tiny depression in the gravelly tan spoil of the footpath. I watch her to find the nest as a couple of walkers go past. She calmly keeps a distance of twenty metres or so between herself and the couple, trusting to the camouflage of the eggs, then returns ten minutes later. Wader eggs are remarkably robust things, and the oystercatchers seem happy enough to leave them lying around.

A lapwing nests in the braids of the Quoich Water, newly created following winter storms and still grassy. It's a high-stakes gamble – the nest is surrounded by water, making it impregnable to predators like stoats and pine martens. But the river is dynamic, and the water level rises and falls with great speed and regularity here. Every time it rains, which is most days this year, I fear the worst. As it turns out, the lapwing knows better than me, and the nest fledges four chicks. Another nest is less fortunate. It is in the main lapwing 'colony', slap bang in the middle of a grassy meadow, kept open by grazing deer and, in late summer, sheep. The camera trap shows a pine marten taking each egg, one at a time. Others in the 'colony' make it through – waders regularly nest in close proximity to one another, a strategy of mutual safety. The more lapwings there are nesting in a colony, the higher the likelihood of each individual nest surviving. Eight miles away, another pair of lapwings gets four chicks away, only for them to be picked off one by one, either by the unseasonal cold or predators unknown. I scan the area for chicks each time I'm passing through, and each time I count one fewer. Finally, I see a single juvenile flying, and chalk it up as a limited blessing.

These are hardy creatures. The cameras show them sitting on eggs, exposed and out in the open, through snow and rainstorms. One lapwing sees off a curious hind, repeatedly. The temperature loggers and nest cameras reveal that 2019 was a relatively good one for lapwings at Mar Lodge, and indeed across the Partnership area, with nests being proportionally more successful than the national average.

Lapwings are cracking birds, and so are oystercatchers, but the real prize is a curlew nest. Up here, they're a summer fixture, moving away to the coast in autumn to get out of the snow. Their whauping,[*] curloo-ing song is a sure sign of spring, and welcome company on

---

[*] Apparently a local name for a curlew is a 'whaup', after their unique call. I've never actually heard anyone calling them that, though, which is a shame. I'd like to bring it back into circulation, but feel it might sound like an affectation in my clipped, southern English accent.

the moors. Through the season I come across a dozen or so curlew territories easily enough – curlews are conspicuous birds when they want to be. There are territories in the marshy bits around the Quoich Flats, in the grassy bits of Dalvorar, and the old pastures of Glen Dee. There are territories on the open moor, and around the edges of the woods. But we run out of time and skill at Mar Lodge to find the nests; curlews can also be exceptionally inconspicuous when they want to be. Others do better: across the ECMP, seventeen curlew nests are found across 2018 and 2019.

Curlew nests are prized by David because they are harder to find than lapwing and oystercatcher nests, but perhaps not as hard to find as redshank, snipe or golden plover. But mostly he prizes them for their increasing rarity. Forget the capercaillie, the eagles, the dotterel and snow buntings, even the hen harriers. The most pressing conservation concern in the UK, according to an RSPB report from 2015, is the curlew.[1] The UK is a major breeding area for curlews, accounting for around a quarter of the global population. But from 1985 to 2015, breeding curlew numbers in the UK dropped by 48 per cent. To put that in a global perspective, we have lost 13.5 per cent of the world population during that period. Over in Ireland, the situation is even worse. Curlews may become extinct as a breeding species in the Republic of Ireland by 2030.[2] Of the world's eight species of curlew, two have almost certainly become extinct within the last fifty years. The reason the BTO is involved in this project is that the East Cairngorms Moorland Partnership area remains a stronghold for declining waders. The idea is to have a good look at what is going on 'on the ground', across a huge area, to work out what specifically is making the place tick. The project is getting stuck in to the complicated, messy business of slowing the seemingly unstoppable decline of an iconic species.

Waders are declining across the UK, I say to David. Why?

'If you look back a century or more, we'd created a heaven on earth for waders. In the lowlands, there was loads of rough, undrained, lightly stocked pasture, which provided them with perfect places to feed and nest. There were thousands of gamekeepers who controlled most of the species that might eat wader eggs and chicks. We also had less woodland back then, which takes up space that waders like lapwings might live in and provides nesting or den sites for many of the species that predate on waders or their nests.

'Then things gradually changed. The intensification of farming in the lowlands drastically reduced the amount of suitable habitat. Lots of potential predators of curlews, like foxes, crows, badgers and pine martens, have increased in numbers. Lowland game shoots are releasing millions of pheasants and partridges into the wild every year, which are providing these opportunist, generalist predators with more than enough food. So lowland wader populations have pretty much collapsed outside of islands and predator-fenced nature reserves. Meanwhile, vast tracts of suitable habitat in the uplands have been lost to non-native tree plantations. And because curlews live for so long, the steep declines we're seeing now may be the result of changes to the landscape that happened many years ago.'

We do know a fair amount about what is causing curlew declines. But a lot of what we know seems to open up more questions than it answers. Take woodland cover. It is often said in Britain that waders hate trees. Studies from the Flow Country, the vast, almost treeless bogs of Sutherland and Caithness, have shown that afforestation with non-native plantations causes not just loss of habitat, but also creates an 'edge effect' – species like golden plover and dunlin are less likely to use areas close to woodland.[3] But waders in Scandinavia are often very strongly associated with woods that have boggy bits in them: wood sandpiper, green sandpiper, whimbrel and indeed curlews can be found breeding within forests. Swedish curlews, which are not declining, are perfectly happy breeding within open, boggy woodland.[4] Indeed, Desmond Nethersole-Thompson, ornithologist extraordinaire to whom we'll return in a later chapter, found curlews regularly breeding in areas of (very) open woodland, 'at the foot of a stunted pine, and exceptionally in the open clearings of the old pine forest'.[5] At Mar Lodge, there are two or three regular curlew breeding sites which sit in what can only be described as extremely open, boggy woodland. Another British study suggests that increasing woodland cover from 0 per cent to 10 per cent of the land area within one kilometre of curlew breeding sites would require a 50 per cent increase in human predator-control effort to maintain population stability.[6] But the two years of landscape-scale studies which we are involved with at Mar Lodge have so far failed to find any correlation between proximity to woodland and reduced breeding success.[7] [8]

Then there's sheep. Grazing sheep in the rough, unimproved grassland of the uplands keeps the grass short enough for waders to use for nesting and feeding. Grazing, be it by wild animals or livestock, is extremely important for wading birds. But in a study of wader productivity in the North Pennines, 20–33 per cent of nest failures were attributed to trampling by livestock.[9] Deer will happily nibble at wader eggs, particularly when there is little else for them to eat.

Then there is the really thorny issue of predation. Like many waders, curlews are long-lived, provided they get the chance to reach adulthood. The oldest-known curlew lived to at least thirty-two years.[10] But the most dangerous time for a curlew is in the nest, and just after it has hatched. So the first thing to remember is that a high level of predation is 'built in' to the breeding ecology of most ground-nesting birds, and the curlew is no exception. Recent work by RSPB researchers has shown that, to maintain a stable population, curlews need to produce 0.5 chicks per pair per year.[11] In a fully functioning ecosystem, this would be fine, but as is often the case, with the breakdown of functioning ecosystems comes knock-on effects. A lack of top predators like eagles, lynx and wolves means more generalist predators like foxes and crows can flourish. Land-use changes brought about by humans can favour predators over prey species, by providing less-than-ideal nesting locations, for example.

Let's not beat about the bush: when it comes to protecting breeding waders, predator control works. If you reduce the amount of creatures which will eat wader chicks, then you will see an increase in wader abundance.[12] But beyond this pretty obvious statement lies an extremely complex web of interactions and unintended consequences. Predator control is far from being a silver bullet for curlew conservation. Take the study which found a strong positive association between curlew numbers and numbers of red grouse and pheasant.[13] This is because the predator control done to protect gamebirds also protects breeding waders. Gamekeepers can put enormous amounts of effort into providing the right conditions for these ground-nesting game species. But this effort also creates a massive food bank for predators, increasing predation levels on birds like curlews. This might explain why the study that found a strong positive association between pheasants, grouse and curlews also found a strong association with crows, which can be a predator of curlew nests. Another paper found that the effect of predator control was highly variable.[14] Yet another study found that predator control on grouse moors does indeed protect breeding waders, but it doesn't need to be done particularly rigorously for the benefits to be seen.[15] In other words, waders will feel the benefits of gamekeepers' work to protect red grouse from predation before the grouse do.

It gets even more complicated. Waders are not the only thing that predators eat. A couple of papers have suggested (though certainly not proved) that sites which hold lots of small mammals like voles might be more likely to have successful wader nests, because there is more food for predators like foxes.[16] Once a certain level of small mammal population has been reached, it might be more efficient for foxes (which, nationwide, are the major predator of curlew nests)† to focus their hunting attention on mammals rather than

---

† Even more confusingly, not a single fox predation event was recorded in the BTO/ECMP study. Foxes are controlled across the ECMP area, including at Mar Lodge.

wader chicks.[17] With some waders like lapwings, population density becomes a factor. The more waders there are nesting in close proximity, the better the overall rate of production of the nesting site.[18]

Which brings us to the limitations of the research done to date. 'There are an awful lot of different interactive factors which correlative research just isn't able to disentangle,' says David. 'In many of the studies "gamekeeper density" is used as the surrogate variable for predator control, but this really doesn't tell us how important it is to control foxes, crows, or mustelids, or how much control is needed for stable populations. And we also don't know how much herbivore [gamebirds, livestock and deer] densities prop up meso-predator populations.'

The plight of the curlew is down to a complex web of factors, but it is easy to get bogged down in the minutiae and lose sight of the big picture. All of this research may appear to be coming to contradictory conclusions about what curlews need from us. But zoom out and it is really telling us something very simple. To survive in the short term, curlews on mainland Britain need the intensive help of humans performing predator control and habitat management in both the uplands and the lowlands. But to survive in the long term?

'This,' says David, 'is where Mar Lodge can be *really* interesting.'

Many species simply aren't suited to modern landscapes. As we have seen with the Kentish glory, such species require edgy, untidy, unprofitable, dynamic landscapes on a large scale. For the curlew, a good proxy for this is provided by the East Cairngorms Moorland Partnership landscape area. In fact, it may even be the case that keepers are doing such a good job that they are creating a 'wader paradise', allowing greater breeding success and higher densities of curlews than you could expect to find in the curlew's 'natural' environment (whatever that might be). Agriculture in the partnership area tends to be relatively low impact, with minimal use of fertilisers and pesticides. Predator control occurs with varying intensity across the area. But the main strength of the East Cairngorms Moorland

Partnership area for waders is its size and its connectivity. Mar Lodge is big, but it is only a small(ish) part of the whole East Cairngorms Moorland Partnership area. The Mar Lodge waders benefit from interacting with a meta-population that stretches out for tens of miles.

But where Mar Lodge brings real value to the project is not in the number of nesting waders it supports (though this is not insubstantial). Rather, it is in the way it differs from other estates. Mar Lodge is not a typical managed upland landscape. The climate is wetter, colder and harsher than the more eastern, low-lying estates of the Partnership area. There is less grazing pressure here, even on the open moors. There is less muirburn. There are fewer developed areas of in-bye pasture. The Quoich Flats, which hold the highest numbers of waders on the estate, are a floodplain. This allows natural hydrological processes to take place, creating amazing ecological niches and helping to protect Braemar from flooding. But it also leaves open the risk of catastrophic nest failure. The Mar Lodge predator control programme is far lighter than on other estates, targeting only foxes and, in some years, crows. It is a landscape in transition, slowly becoming more like the wader haunts of Scandinavia than the grouse moors of eastern Scotland.

The moors of the East Cairngorms Moorland Partnership provide a tried-and-tested formula for curlews. Here, lessons can be learned about landscape-scale wader successes. Mar Lodge is at the experimental edge of that landscape, the wilder end, where new knowledge is to be found. The Mar Lodge ethos is to create the conditions in which wildlife can look after itself, and not to be too concerned about having management targets for particular species, except where direct intervention is needed to conserve species of national and international importance. Mar Lodge possesses all the building blocks of 'natural' curlew habitat – a dynamic mosaic of wetlands, subject to the natural processes of grazing and flooding. There are areas with higher levels of grazing and lower levels of grazing. There are dense patches of rushes for nesting in, and close-cropped grazed land for feeding in. The big storms of 2014 and 2016 supercharged wader breeding on the Quoich Flats by creating new, permanent wet areas, braiding around the river. They did a better job of managing the land for waders than we could have hoped to have done ourselves. But it may be that those benefits subside over successive years, until the next big storm comes along.

Curlew species across the world lived perfectly well without human intervention for millions of years, but in the last century they have struggled to adapt to fast-changing modern landscapes. The East Cairngorms Moorland Partnership provides them with a huge, joined-up area of managed landscape, in which they appear to be thriving. Mar Lodge aspires to a different kind of landscape, which will have both positive and negative impacts on wader populations. It makes Mar Lodge a landscape-scale laboratory in wader conservation. Taking such a step involves changing the conservation mindset, learning to let go of targets and plans and trusting to nature. It's terrifying, but it provides amazing new insights into the workings of our landscape and ecology. Such is the fun of cutting-edge conservation – it pushes the limits of what we know.

What if, for example, all our nest-finding efforts show us that the Quoich Flats are a population sink, where waders struggle to breed successfully? If it is, then maybe we could think about better, more intensive ways to manage it for waders. Or maybe we could allow it to revert to willow scrub – a really rare habitat in the uplands which could provide a home for other endangered species like whinchats and redpolls,‡ while encouraging waders to make fuller use of better habitats elsewhere. But what if curlews *are* doing particularly well here? How could we use that information to inform other curlew conservation projects?

What if curlews are breeding at low densities across the breadth

---

‡ Or potentially even rarer species, which have become extinct across most of Scotland but could, potentially make a comeback: cranes, perhaps? Bluethroats? Beavers?

of the estate, successfully using the landscape in response to newly created niches and varying predator abundance? What if the curlew can be a bird of the woodland edge, like black grouse, but we have simply forgotten that this is the case?

'One possible path for curlews is that they contract their range to the islands and bogs and breed at really low densities, maybe a few thousand pairs across the Highlands, like greenshanks,' says David.

How will we respond to that eventuality at Mar Lodge? What if, sometime in the future, driven grouse shooting is banned, or heavily licensed, and grouse moors across Scotland become less intensively managed? With its low-intensity moorland management, Mar Lodge again could show what the future holds.

There's also an element of hedge-betting at play here. When species come to be dependent on a single land-management practice, this can leave them in an imperilled position should political or social circumstances change. If we can show that curlews can be successful in other landscapes, so much the better. 'For many years,' says David, 'our curlews have been dependent on a cultural landscape which might not persist for that much longer – if predator control, livestock grazing and muirburn decline in the uplands while woodland cover expands rapidly, after a while our curlew population might look very different. Are we happy to have a few thousand pairs scattered here and there in remote Highland glens – or are we going to try to hold on to as many as we can? Maybe this isn't a question for nest finders or complicated models – it's more a question of how much we value this bird, how we want these hills to look, and who or what is this landscape?'

In July, I'm delighted to see two spindly, ridiculous creatures skulking through the rushes near one of our monitored curlew territories. Curlew chicks may be rare, but they are certainly not the most beautiful of creatures. A week later, two more appear feeding in short sward a mile away. Two more turn up at the Quoich Flats. Another one appears on a lonely patch of bog in the regeneration zone. Something is working. Time is running out for the curlew across huge swathes of its British range. But Mar Lodge is part of a larger landscape where curlews are proven to be surviving, and surviving well, giving the species and those who would save it the luxury of space, time and data.

11

Hen Harrier

In five Aberdeenshire parishes, clustering around Braemar, 70 Eagles were slain in the ten years from 1776 onwards.

James Ritchie, *The Influence of Man on Animal Life in Scotland*, 1920

It is disappointing that, again, there continues to be persecution of birds of prey.

Roseanna Cunningham MSP, Cabinet Secretary for Environment, Climate Change and Land Reform, 'Wildlife Crime in Scotland: 2018 Annual Report'

I pick her up out of the corner of my eye, flying determinedly back to her nest. She's half a kilometre further east than last year. Cursing, I run around the deep gulley, a flurry of optics and thermal layers, but I'm too late. She's back down, and so I'll be here for a while longer. The mizzle sets in, I pull on the waterproofs, sink further into the natural stone shelter I've found, and protect my soggy notes as best I can. My binoculars steam up. The study of birds of prey, working out what they're up to and how we can help the species in trouble, is one long jigsaw puzzle where the picture on the box keeps changing and the pieces keep getting wet.

An hour later and I'm out of coffee, but the mizzle has cleared a bit and there's a vague brightness in the sky. I stretch, not looking away from the spot where I think she is. To the right, a grey ghost appears out of the murk, beating in on paper-light wings. Silver with ink-black wingtips, he's calling excitedly. He's too far away to pick them out, but he has fierce yellow eyes. He's got something trailing in his talons. As I strain to see what it is, he is joined by the female, a larger, brown bird with a distinctive patch of white on her rump. He drops his parcel, a vole. She plucks it out of the sky, while I do my best to ignore his calls and put my full attention onto her. She drops out of view, behind yet another gulley, but miraculously comes out again, and drops like a stone into a particularly deep, thick patch of heather. I've found the nest. I take pictures and sketches, but the rain is coming in, so I retreat back to the office.

I return that evening when the rain has stopped and the heather has dried out. It's a steep climb – the Mar Lodge harriers nest at a higher altitude than almost any other harriers in the country – through thick vegetation. It's demanding work, and I've lost a stone in weight since January. I match up the features of the hillside with the pictures and sketches, calling to her, 'Come on girl, up you come.' Finally she explodes out the heather, two metres away, and she's furious. Four white eggs, the spitting image of the ones you have for breakfast. This is a privileged view – it is illegal to visit a harrier nest site unless you have a licence. She dive-bombs me, a warning shot, pulling up a couple of metres above my head. I take the grid reference and altitude and a couple of photos and leave. Two minutes later, she's back on the nest.

Later in May, and I'm still not certain whether there is just the one nest here. Hen harriers are polygynous; sometimes females share a male, and this might be the case here. I sit under a tree on the other side of the valley, waiting for the rain to stop. This year, May has been a miserable month of drizzle and snow. The female comes off the nest. She's not back for an hour – surely that's too long? Three hours later, the male's a no-show, and I'm feeling a bit funny about this one. I'm expecting it to fail. I'm loath to visit the nest site any more than is absolutely necessary, so I keep my thoughts to myself and head back out of the glen, deflated. But I'm certain now that there is only one nest here, so the morning's job is done, another piece of the puzzle is filled in.

I visit the nest a couple of weeks later for the second of three visits, and I'm surprised how far behind schedule it is. The chicks are only just hatched, but they're alive all right. I'm over the moon. The snow and rain of early May has pushed things back for some of the birds, but not all of them.

I make my third and final visit in June. The steep face has lost none of its potency. I startle an adder as I get close to the nest site but ignore it – I want to be quick. I spy a flash of white in a suspicious spot. 'No no no,' I'm muttering to myself between heavy breaths, running as fast as the heather will allow. On an old deer track, a mass of feathers sprawl out, white, greys and blacks. The male has met a sticky end. The female hasn't flushed, and I see why. She's dead too, next to the nest, a mass of beetles and maggots throbbing in what used to be her breast. The only consolation is that they met a 'natural death' at the talons of a larger raptor – a goshawk, maybe, or an eagle. The wet weather and a population crash of voles has made this a hard year for all predators, and they are going to more drastic lengths to feed their young. The bodies of two of the chicks are still in the nest, pathetic blobs of pink against the black-brown nest. I swear a bit, kick a patch of heather, take the samples and photos I need, and start on the long trudge back down the glen. Another piece of the puzzle filled in.

We have other, happier stories in 2019. We first heard of one particular harrier nest from Barney, an extremely pleasant chap doing fieldwork for an MSc. Barney's the sort of person who happily surfs fifteen-foot waves* in Shetland's freezing waters, and then doesn't tell anyone about it.

As he was walking through the moors he saw a female harrier dropping into deep heather. Hen harriers are a Schedule 1 species,

---

* Like anglers, surfers are also obstinately imperial in their outlook.

making it illegal to disturb them at the nest unless you have a licence to do so. Barney therefore didn't go any closer to investigate, but reckoned he had the bird's location fixed to within an area of about three hectares. In Mar Lodge terms, where there is about 10,000 hectares of suitable heather to choose from, Barney's information was gold dust.

Finding bird-of-prey nests, I should say, is a bit of an art. It is also extremely frustrating. A three-hour vigil leads to nothing more than a wasted morning with alarming regularity. A single lapse of concentration for half a minute can render a whole morning's work redundant. In Scotland, it is cold and often wet work. It involves repeatedly trudging up godforsaken hills, sitting very still for several hours and waiting for something to happen a fair distance away. And when the thing does happen it is usually over in less than a minute.

I go out the next morning, (another early morning, another existential crisis), and watch Barney's site from a kilometre away. Three hours in and I'm due for another appointment. I will have to come back and sit them out. I turn around to sketch the ground and take a photo of the rough location. And there it is, that instant, a food pass. I scramble for the telescope, sending my notepad flying. The male, gorgeous, slate grey and white with black wingtips, clear as day, dropping a gift to his mate. She plucks it from the air and disappears into the heather. I simultaneously whoop and curse my stupidity and serendipity. I fix the image in my head, then take more photos, and then head off for breakfast and work.

The next evening, in the next weather window, I stalk up the hill. It is all a rather breathless affair. It's another rough location, the heather is getting on for a metre deep, and the hill is steep. The bird shoots out, slightly further right than I expected, and immediately starts chittering at me. Harriers normally only flush from the nest when you are within a couple of metres of them, but this one jumps off when I'm a bit further off. I rake into where I think the nest is, and a nervous minute later I find it. Harrier nests are scrappy affairs, no more than a mess of beaten down heather, vole carcasses and droppings. And there it is, a single chick, maybe a week old, fat and plump and grotesque and furious, glaring with fierce glowing eyes, lying on its back, talons raised, waiting to rip out my eyes if it gets the chance. The female is chittering above my head, diving on me, pulling away at the last moment. I take a photo and a grid reference and walk away. A few hundred metres away I hear another tone, the male call, less urgent than the female's angry chitter. I turn around, see another food pass, and down she goes, back onto the nest, and that's that.

Four weeks later, I revisit Barney's nest with Brian Etheridge. Brian is part of the Scottish Raptor Study Group, a remarkable, volunteer-led organisation of dedicated, highly skilled ornithologists who provide the majority of our information on the state of birds of prey in Scotland. The most remarkable thing about them is that they provide their ornithological services for free, doing a job that the government would otherwise have to do, saving the taxpayer an estimated £1.8 million every year.[1] The group are an integral part of the raptor monitoring work that goes on across the estate, with their species of interest including merlins, peregrines, eagles and 'honorary raptors' ravens. One of their group, Graham Rebecca, has done a lot to teach the Trust staff on the ground how to go about monitoring birds of prey.

Brian is one of only a very select few people licensed to fit satellite tags to birds of prey in Scotland. Far from the breathless affair of finding the damn nest, Brian works with an efficiency that can only come by forgetting for a moment that one is working with supremely precious creatures. However, like many things in conservation, bird ringing and tagging is a mixture of extreme skill and technology combined with more than a small element of farce. The chick is within a day or two of fledging, and could potentially fly the nest, so Brian has his own system of ensuring that chicks do not scarper when he comes to tag them. Basically, it involves sneaking up on a nest as slyly as possible, and then at the last moment throwing an old fleece over the chicks ensconced within. It works very well: the fleece is heavy enough and dark enough to keep the chicks calm, light enough not to harm them in any way, and cheap enough not to worry about when it inevitably gets covered in harrier droppings. We checked the size, weight and sex of the chick. We ring him with a standard BTO ring and a special colour-coded one. Brian is dextrous, with good reason – harrier chicks are angry and armed with razor-sharp talons, as I find out when the chick takes a swipe at my thumb. Brian takes his time, carefully sewing the harness that carries the tag around the bird's breast, measuring the weight of the ribbon. Everything in his box has a correct place: here is a box of swabs for DNA testing, here is an envelope to collect the spent ribbon, each marked with a special coding system. My job is to scribe, photograph and distract the enraged parent female.

Brian fits the tag lovingly, talking to the bird as he goes along. The tag has been paid for by the EU, who have funded an RSPB-led, five-year project to conserve hen harriers.[2] It is truly tiny, an incredible piece of kit, solar powered, very reliable, able to locate the bird to within a hundred metres or so at regular intervals through the day. By law, it is less than 3 per cent of the body weight of the bird. Extensive testing has shown that they have no impact on the bird's ability to live unhindered.

All tagged chicks need names. It is rather a privilege to name a tagged bird: there aren't many of them around. I name him Ingmar.[†]

Hen harriers are less well known than other birds of prey, but the people who do know them regularly fall in love with them. They fly in a weightless sort of way that means you half expect them to be

---

† You're probably wondering about the name. By 2018, we had so many harriers tagged at Mar Lodge that it was a struggle to remember which ones were related to which. We also wanted a quick way to keep track of individual birds in the news. So we came across a plan that each bird would have 'Mar' somewhere in the name, with the placing of 'mar' in the name relating to where on the estate the bird was reared. By 2019, we were running out of names.

blown over the next hill by a small gust of wind. Rather than the straight line, fighter-jet-style speed of a peregrine, they move with the aerial prowess of an attack helicopter, raining death from above on voles and small birds. In fact, it is their remarkable capacity for hunting things that are on the ground which has led them into a spot of bother or two over the last couple of centuries.

Back in 2016, a pair of hen harriers bred at Mar Lodge for the first time in living memory.[3] In 2017, they returned, and were again successful. Then, in 2018, there were seven nests, of which six (six!) successfully reared young. It coincided with a boom in vole numbers. Twenty-four birds fledged; a ridiculous amount. At one point in 2018 we had over thirty of one of Britain's rarest raptors flying around the glens. It was an interesting time for the raptor workers scrambling to locate the nests. Hen harrier nests are notoriously difficult to pin down. One nest, in a particularly challenging spot, took a full week of watching to find. Another took just ten minutes. One was found by a walker who got lost on his way up a Munro. In 2019, another seven females attempted to nest, but following the atrocious weather and the crash in vole numbers, only three nests successfully fledged birds.

There is much speculation as to why harriers are suddenly thriving here when they have been all but lost as a breeding bird from the rest of Aberdeenshire. It was often said that Mar Lodge was too high and too cold to support harriers. This has proved to be categorically incorrect. Reduced grazing and burning at Mar Lodge has created large areas of long heather, which is great for nesting in, and may support higher numbers of voles than shorter heather. Harriers are spending more time in regenerating woodland than we might have expected. It is a popular misconception that grouse moors are the hen harriers' preferred habitat. For one thing, grouse moors as we know them have only been around for just over a century, and the birds had to nest in other places before then. For another, their stronghold in the UK is Orkney and the West Highlands, which don't have any grouse moors. While they definitely like to hunt open ground, we are beginning to think of the Mar Lodge birds as being equally at home around the edges of the woodland as they are on the open moorland.

Our knowledge from our ground studies has been massively increased by the satellite tag data. We can see where 'our birds' disperse to in winter (the Aberdeenshire coast is a hotspot, but birds have travelled as far south as Yorkshire) and how many birds return to breed at Mar Lodge. We can see how far they can travel in a day (up to a hundred miles or so), and how loyal they are to their favourite sites (it's an individual choice). It's shown us what land they are hunting over (moorland and farmland mostly), and once or twice it has helped pinpoint nest sites for us.

The main reason that harriers are thriving at Mar Lodge is that we don't kill them. It is an unequivocal truth, if not one that is yet universally accepted, that driven grouse shooting is categorically linked to widespread illegal persecution of birds of prey. Despite being made illegal in 1954, every year scores of harriers are killed on moorland which is managed principally to support the sport of driven grouse shooting.‡ This is because on intensively managed grouse moors the easiest things to eat are the overabundance of young grouse, and harriers are exceptionally good at doing just that. The problem is, the harriers are eating the young grouse that moorland managers want their clients to pay to shoot later in the year.

When a bird with a satellite tag dies in normal circumstances, the RSPB staff monitoring the tag see a highly distinctive, slow decline in the quality of the data, followed by a geographic cluster of data points and a slow loss of power in the tag. They can then use this data to find the bird and perform an autopsy. When a tagged bird dies in suspicious circumstances (in the parlance, 'mysteriously disappears'), there is none of this. There is simply an abrupt radio silence. The tag simply and suddenly stops transmitting data.

The first Mar Lodge bird to 'mysteriously disappear' was Calluna.

‡ Some, but by no means all, driven grouse moors.

She was tagged in June 2017. Just six weeks later, on 12 August, the Glorious Twelfth, the first day of the grouse-shooting season, she 'mysteriously disappeared' over an Aberdeenshire grouse moor. Next to go were Margot and Stelmaria, who disappeared over Scottish grouse moors in August 2018, just a month after they left Mar Lodge ground. Of the birds tagged in 2019, two more birds, Marci and Romario, were lost in mysterious circumstances in Yorkshire and the Cairngorms National Park, respectively. In 2020, another bird disappeared, close to where Romario had disappeared the year before. At the time of writing, over a third of the birds fitted with satellite tags at Mar Lodge have 'mysteriously disappeared' over grouse moors. From January 2018 to January 2020, across the British Isles thirty-one harriers either 'mysteriously disappeared' or were proven to have been killed by grouse-shooting interests. This includes birds which were caught in illegal traps, shot and poisoned.

Despite being host to dozens of tagged birds in the last few years, several of which have died of natural causes on the estate (and which were recovered and had autopsies), no tagged birds have ever 'mysteriously disappeared' at Mar Lodge. From 2016 to 2019, eleven nests from sixteen confirmed breeding attempts successfully reared forty hen harrier chicks at Mar Lodge, and no birds 'mysteriously disappeared' on Mar Lodge land. In the same time period, all of the driven grouse moors in Deeside combined raised zero harrier chicks, while eight satellite-tagged birds 'mysteriously disappeared' over their land.

Raptor workers, the dedicated individuals, usually volunteers, who study our birds of prey, always suspected that illegal persecution was widespread. But raptor persecution occurs in lonely, isolated glens, often under the cover of darkness. It is extremely difficult to catch someone in the act. From time to time, raptor workers would find a bird that had been poisoned, shot or trapped. More often they would find a nest site abandoned for no good reason. Occasionally a keeper would be caught in the act and given a slap on the wrist in the courts.[4][5] But it wasn't until conservation groups started fitting satellite tags to birds that the eye-watering scale of criminal activity underway over Scotland's moors was proven beyond doubt. A study of satellite tag data from Scottish golden eagles, including Mar Lodge birds, found that up to a third were 'mysteriously disappearing' over grouse moors.[6] A recent study from England found that 72 per cent of satellite-tagged hen harriers were suspected to have been illegally killed. Birds were ten times more likely to 'mysteriously disappear' over grouse moors than they were over any other type of land, per unit of time spent on each type of land.[7] The RSPB has a database of 558 confirmed cases of raptor persecution in the UK from 2002 to 2017.[8] From March 2017 to March 2018, Police Scotland recorded twenty-four instances of crime taking place against birds of prey.[9] These figures do not include a further eight satellite-tagged birds of prey, including Mar Lodge birds, which 'mysteriously disappeared' in the same period. This is above the five-year average of raptor crimes. Contrary to what you may hear in press releases from grouse-shooting lobby groups,[10] and seventy years after the practice was made illegal, raptor persecution is not going away.

Across Scotland, generally speaking, birds of prey are doing better than they have for centuries. Ospreys have returned after being persecuted to extinction. White-tailed eagles, also once driven to extinction, have reached the 100-pair mark and are now breeding across the breadth of the country. Goshawks, also once persecuted to extinction in the UK, have now returned, thanks in no small part to the actions of falconers. Red kites are flourishing. Peregrines and buzzards have bounced back, thanks to the banning of DDT and other harmful pesticides in the mid twentieth century, which nearly drove them to extinction as well. Of all our larger raptors, the hen harrier is the only one that is declining, and it is declining fast, and we know that the reason hen harriers are declining is because gamekeepers are illegally killing them.

There is grouse shooting, and there is driven grouse shooting. Grouse shooting (either 'walked up' or 'over pointers') is something that is done all over the world, wherever there are species of grouse to shoot. This is the walked-up sport that takes place at Mar Lodge. Driven grouse shooting is another beast entirely. Driven grouse shooting is unique to Britain, having been developed in the nineteenth century. A line of 'beaters' walk across the moor towards a line of 'butts' (essentially fancy trenches). The beaters scare all of the grouse into flying towards the waiting shooters ('guns'). It is good sport because grouse fly very quickly and not very far above the heads of the waiting guns. The difference in scale between the ancient sport of walked-up shooting and the 'new' sport of driven grouse shooting is well expressed by the Mar Lodge gamebag. In an average year, the same amount of grouse shot at Mar Lodge over the course of an entire season might be shot in a single day on a single beat on an intensively managed grouse moor.[11] The majority of the 120 or so grouse moors in Scotland operate as driven grouse moors.

The problem lies in the fact that driven grouse shooting involves producing a hyper-abundance of grouse to shoot at – far more than would ever naturally occur in the landscape. Once you have created a very large, shelterless area containing a single prey species which you want to kill yourself in large numbers, you then need to kill absolutely anything that might even think vaguely of eating them. In these hyper-managed artificial environments, a single hen harrier can do a significant amount of damage to the balance sheet of a sporting estate, particularly if grouse shooting makes up the majority of the income of that estate. Put simply, hen harriers are being illegally killed on some moors so that people can kill red grouse for sport, and then other people can make money from it. And it's not just harriers: eagles, peregrines, goshawks and buzzards are all regularly killed illegally across Scotland's grouse moors.

Mar Lodge has always been marginal ground for driven grouse shooting. With a few exceptions, it is either too high, or too wet, or both, to support the numbers of grouse needed for successful, regular driven shooting. So marginal is most of the land here for grouse that to regularly drive birds would take so much intensive management that it would be both uneconomical and hugely environmentally damaging. But it is excellent ground for low-intensity, walked-up shooting. Beautiful, remote, the challenging terrain provides a great test for the shooters' skills. It is a place where grouse shooting and hen harriers get along perfectly well together.

Working with the Mar Lodge hen harriers is exciting, and a real privilege. Through quirks of history, changing land management and ecology, we are witnessing the return of a predator in good numbers to an evolving, regenerating landscape. It is expanding our knowledge of what the bird is capable of and how it responds to new landscapes. In April, the skies come alive with the undulating aerial mating displays of harriers, their famous sky-dancing antics. But every spring also brings anxiety. Will they return this year? How many of our birds will have been shot?

Deeside has an ignominious reputation for bird-of-prey persecution, and the practice casts a long shadow over an area prized for its natural beauty and royal associations. By the late nineteenth century, Deeside's goshawks and white-tailed eagles were gone, and buzzards, common now, were also extremely scarce on the ground. Honey buzzards were shot out in 1867 and 1893. Red kites were exterminated.[12] In the period from 1933 to 1938 the influential naturalist Desmond Nethersole-Thompson came across nine eagles killed by gamekeepers in the Cairngorms. In one instance, an egg collector lifted the eggs of a nest from below the corpse of an eagle. It was, in his words, 'all very naughty'.[13] Even back then, it seems that Mar Lodge birds may have gotten off lightly: Adam Watson records the actions of the famous, much-loved gamekeeper and friend to mountaineers, Bob Scott. While never particularly keen on killing the golden eagles on his beat, he gave up the practice entirely after the RSPB introduced a £10 bounty for every successful nest. His reticence to shoot eagles led him to suffer the ire of other local keepers.[14]

This is by way of saying that old habits die hard. There is a long, long history of raptor persecution in Scotland, and since sport shooting became the main social driving force in Deeside, it has also been the driver for raptor persecution. Raptor persecution is ingrained. But it needn't be this way forever.

The tide might just be turning for Scotland's harriers. The information that satellite tagging has yielded has proved to be a game changer. More people than ever are aware of the plight of the hen harrier, a creature that few had heard of until just a couple of years ago. This is thanks to a decade of hard campaigning by the likes of Ruth Tingay's Raptor Persecution UK, ex-RSPB director Mark Avery, and broadcaster Chris Packham. New tech, not just satellite tags but also camera monitoring kit, ubiquitous smartphones and social media, are tightening the screws on the injustices inflicted on our raptors. Scottish government is making noises about regulating sport shooting through a licensing scheme, a move long campaigned for by RSPB and long vehemently campaigned against by grouse-moor owners and their representatives. Even the climate seems to be turning against driven grouse shooting. Extreme heat in 2018, followed by unusually cold, wet weather in the spring of 2019 led to many shoots cancelling driven days, due to a lack of grouse.[15] These years were also poor years for the Mar Lodge grouse. Ultimately, it may be that driven grouse shooting is finished not by environmental campaigners, but by climate change.

On 29 January 2020 a partnership of influential 'countryside associations', including the British Association of Shooting and Conservation, the Countryside Alliance and the Moorland Association, released a statement on raptor persecution. It firmly denounced those that persecute raptors, calling them a threat to the future of country sports.[16] They will, they say, be introducing a range of measures designed to stamp out raptor persecution. Sixty-six years after raptor persecution was made illegal, it is their strongest ever action on the subject, but it is a rearguard action. They know that raptor persecution, something that was plausibly denied by grouse-moor interests for decades and was once only known about by a small number of raptor enthusiasts and conservationists, now threatens to destroy the entire shooting industry. In a video released as part of the same PR blitz, Eoghan Cameron, chairman of British Association for Shooting and Conservation, says:

> They are doing the rest of us a severe disservice, they are prioritising themselves above all others, and they are sabotaging shooting. I think it's about time that we looked at those who do persecute raptors through the same lens as sabs, they're no better, and quite frankly they're handing out ammunition to those that would see us stop shooting. So if you know of anyone who is doing it, it needs to stop. It needs to stop now, otherwise sport shooting will suffer.[17]

Then again, things might not be changing after all. In the spring of 2020, as you and I were locked down, sacrificing our freedoms to fight Covid-19, certain gamekeepers across the country were using the time to kill as many birds of prey as they could.[18] Yet another Mar Lodge harrier 'mysteriously disappeared' during this time.

Odd as it seems, Mar Lodge's harriers may just be the salve that Highland grouse shooting needs to survive. Our harriers have helped to highlight the robbery of our shared natural heritage for the benefit of those who would overexploit red grouse, another part of our shared natural heritage, for the sake of profit and tradition. But they are also proving that grouse shooting *can* coexist with harriers. More than that, they are showing that harriers can thrive alongside Highland sport. Mar Lodge's keepers are delivering walked-up grouse shooting alongside a growing harrier population.

This is one potential vision for the future of Scotland's grouse moors. The Mar Lodge shooting model aspires towards lower bags of more quarry species at a higher price, with the profits ploughed back into managing a dynamic landscape where blanket bogs intermingle with heather moor, willow and birch scrub, with alders, aspens and willows along the river banks and a natural transition between the open moor and the Caledonian pinewoods. Highland sport will be creating homes not just for red grouse, but for any number of our quarry species.

It's 2070. A group of sports folk heads out for a day on the moors. Their ghillie, dressed in the distinctive, salmon-pink Mar Lodge tweed, reads them the quarry list for the day, but they are distracted by sand martins and swallows, too many to count, drawn in by the insects breaking rank from low willow scrub. A fat trout plops out of the water, jumping for insects, while a small group of stags breaks the skyline, their antlers silhouetted against the clouds. The hunting group emerges from the wooded river valley, spaniels loping ahead of them, into a mosaic of heather and dwarf birch. They are surrounded by snipe, and a covey of black grouse flies overhead while a late skylark sings sweetly, its voice intermingling with calling dunlin, robins, willow warblers, tree pipits. A plump red grouse flushes; there's the distinctive report of a shotgun, and the shooter's dinner is sorted for the evening. Floating above the shooting party, the silvery enigma, the guardian of a healthy moor, is the hen harrier.

Part Three
IN THE MOUNTAINS

February can be a savage month in the Cairngorms. It has the most snow and ice, the most avalanches, the lowest temperatures, and the most fatalities of any month in the mountains.[1] It is also among the most beautiful months, with icing-sugar coated mountains, pancakes of ice swirling in the eddies of fast-flowing burns, pines draped in a blanket of snow, pure white mountain hares, golden eagles displaying their prowess to potential partners with death-defying stoops, banks and rolls, and, from time to time, the merry dancers in the sky, the northern lights.

February 2019, following a particularly vicious blizzard, a stable ridge of high pressure settles over the Cairngorms. This means bright, clear blue skies and bitterly cold temperatures. Braemar makes the national news for being cold, reaching −15.5°C. Checking the forecast, Shaila and I decide to blow off work for a day and go hillwalking. We choose the cone-shaped Derry Cairngorm and Sputain Dearg, a subsidiary top of Ben Macdui which sits above the particularly spectacular Coire an Lochain Uaine. The avalanche forecast suggests that snow conditions would be good, and not dangerous, and we figure that the walk could be comfortably accomplished in a long day.

We set off early, over a frozen Derry Burn, the snow crunching and squeaking. It's good snow: the temperature has remained below freezing since it has fallen, so there has been no time for it to melt and refreeze into ice. We admire the 'fine firs', as Queen Victoria had done, and marvel at the regeneration marching up the hillside. Sorrel the dog trots alongside, playing with sticks.

There is no crust to the snow to support our weight, so we take it in turns to break the path. It is hard work, but somewhat pleasurable if you are inclined towards masochism. On reaching the plateau at Creag Bad an t-Seabhaig, the peregrine crag, we are joined by the croaking of ptarmigan and small flocks of snow buntings. We admire the cornices around Sputain Dearg, the frozen lochs Uaine and Etchachan. We spend the day chatting and admiring the view, which stretches from Ben Nevis in the west to Peterhead in the east. We photograph the corries, the great, complex vista that stretches from Braeriach to Beinn Bhrotain. We meet a chap on Sputain Dearg, also pinching himself that he has happened to find himself in such a paradise. He is in the hills for a couple of days, camping and, if the weather turns rough, retreating to the shelter of the bothies. We taunt ourselves with vertigo by looking over the precipices, seeing where the cornices have collapsed and avalanches have fallen. On the way down, we surf the snow in waterproof trousers, steering and stopping with ice axes. When the sun goes down, slogging through the thick powder of Luibeg, my face starts freezing. We startle a covey of red grouse and flocks of bullfinches feeding on heather seeds. It is the sort of day that walkers in Scotland expect to get maybe once a year.

The next day, the weather breaks and conditions in the mountains turn into a hellish mixture of blizzard, white-out and wind. Such is the fickle nature of the Cairngorms winter.

It turns out that, as we were descending Sputain Dearg, a photographer picked us out from the summit of Derry Cairngorm, well over a mile away. We appear in the picture he took as two dots above a mass of snow and rock, silhouetted against the pale blue sky. By contrasting the scale of ourselves with the enormity of the mountain, the photo magnifies our meagre efforts: we, mere dots, take on the appearance of heroic adventurers, pitting ourselves against a sublime, hostile landscape. With the ego of an aspirant mountaineer, I'm having the picture framed.

If there is one thing that the Cairngorms are renowned for in Britain more than anything else, it is space. This is a big place. The Ordnance Survey recently determined the longest straight-line distance that you can walk in Britain without hitting a road. Unsurprisingly, they came up with a line that goes straight through the Cairngorms. And that 45-mile line goes slap bang through the middle of Mar Lodge Estate.[2]

Starting at Corgarff, famous for its castle, the line hits Mar Lodge territory at the summit of Beinn a' Bhùird. It cuts south-west across Beinn Bhreac, Carn Crom, Carn a' Mhaim and Beinn Bhrotain, before hitting the march with Glenfeshie on the Mòine Mhòr, the great moss, cutting through the lonely expanse of Gaick and finally hitting the A9 north of Blair Atholl, somewhere around the middle of nowhere. To walk the line would be some undertaking. It would involve scaling the corries of Beinn a' Bhùird and taking on all the extensive downs and ups through Glen Derry, Glen Luibeg and Glen Dee.

There are two interesting things about this line. The first thing is how long it is: a true traverse of it would take about three days and would involve some fairly gnarly rock climbing. It's nice to think that you can still walk for three days in Britain without hitting a road. The second thing that is interesting about it is how short the line is. Only three days? The largest tract of roadless land in Britain? Is that it?

This contradiction lies at the heart of understanding the Cairngorms. If you can get your head around the fact that the massif is both eye-splittingly large and painfully small, then you can understand the immense pressure that we are putting the landscape under. It is a wilderness hemmed in by civilisation, a refuge imprisoned by development. There is an irony in the fact that the least disturbed, most 'natural' ecosystem in the UK is also among the least biodiverse. But then, it is the sheer savagery of the Cairngorms, the cold, the wind, the snow and ice, the driving rain, the complete and total lack of shelter that renders it deadly for all but the most resilient, best-adapted creatures, that has also protected it from human destruction.

It is exactly this savagery, the fierce beauty that the Romantic poets and artists called the 'sublime', that draws people to the mountains in their thousands. The high tops of the Cairngorms are, more than any other place in the UK, wild. Their lure is their remoteness, their relative ecological intactness (purity, if you will), their beauty. They are attractive to us, ultimately, because they are as far from the sanitised, air-conditioned, stale, unhealthy, comfortable office environment as you can get in the UK. And yet they are also accessible in a couple of hours from Edinburgh. Given a fair wind, a spell of good weather and a long summer day, a reasonably fit, well-prepared person can climb Ben Macdui, Britain's second-highest mountain, in a day. The Cairngorms offer a place where people can play at having grand adventures, and still be home in time for a takeaway and *Strictly Come Dancing*.

And yet Scotland's mountains can be a deadly place. The Cairngorms are responsible for some of the worst mountain disasters in the UK. For many visitors, the frisson of danger and death is deeply attractive. This is a theme picked up by Robert Macfarlane, who writes of the peculiar fascination that humans hold for the mountains, and how this fascination has evolved over the centuries, from horror to aesthetic appreciation to playground: 'Mountains seem to answer an increasing imaginative need in the West,'[3] he writes. They 'challenge our complacent conviction that the world has been made for humans by humans'. The ultimate paradox is that many people feel at home in the mountains in the way that they do not in a city.

The mountains, then, are a dangerous, inhospitable place, where people go to challenge themselves, to scare themselves, to prove themselves. But Nan Shepherd, the greatest writer of the Cairngorms, wrote of walking in the hills as 'visiting a friend'.[4] For her, there was no glory to be had in scaling the peaks. For Shepherd, the greatest prize in visiting the Cairngorms was not the summit cairns, the tricky climbing lines, the totems of human mastery over the mountains. Rather, it was the chance to go 'into' the mountain. Far from being an inhospitable, dangerous testing ground of strength and resilience, or a wilderness in which walkers can 'find themselves,' or a backdrop for an Instagram story, for Shepherd the lure of the mountains was the possibility afforded to become a part of something greater than mere humanity, to notice the interconnectedness of everything from the rocks to the moss to the flowers, to the snow to the animals to the humans. For Shepherd, the chance to be a small part of a 'living mountain' was an infinitely greater prize than being a mere human conqueror of a landscape.

For all that, Shepherd would ultimately have agreed with Macfarlane's pithiest observation: 'Those who travel to the mountain tops are half in love with themselves, and half in love with oblivion.'[5]

Mountains may harbour some of the last refuges of wilderness in Britain, but they are extremely easily disturbed and destroyed by humans. The shallow and non-existent soils and the intolerable climate make plant growth painfully slow, and the creatures reasonably sparse, and the ecosystem reasonably simple. This makes mountain ecosystems hugely susceptible to damage by humans, however well-meaning they are. Through the mere act of visiting the mountains, humans can very easily destroy the very qualities that lured them there in the first place. For this reason, the work of conservationists in the mountains is often not the management of the land, but rather the management of the people who visit it.

Such an observation is nothing new. The complicated, tautological business of 'managing' wild mountains has been a philosophical struggle that the National Trust for Scotland has wrestled with since its acquisition of Glencoe in 1937. Percy Unna, the influential mountaineer who helped facilitate the acquisition, wrote down a list of prescriptions for the future management of Glencoe. The 'Unna Principles' broadly argued that mountains should remain in a state of what we now call 'wild land'. They included stipulations that 'the hills should not be made easier or safer to climb', and that 'no facilities should be introduced for mechanical transport; that paths should not be extended or improved; and that new paths should not be made'. Indeed, Unna was keen to make visiting the mountains as uncomfortable and difficult as possible. Two other principles were:

> That no directional or other signs, whether signposts, paint marks, cairns, or of any other kind whatever, should be allowed; with the exception of such signs as may be necessary to indicate that the land is the property of the Trust, and to give effect to the requirement in the Provisional Order of 1935 that By-Laws must be exhibited.

> That no other facilities should be afforded for obtaining Lodging, Shelter, Food or Drink; and, especially, that no Shelters of any kind be built on the hills.[6]

The Unna Principles are among the most influential documents regarding how Scotland sees both its mountain areas and itself in relation to them. They rely on the perception of the mountains as a place inviolate, and the perception that wildness is a necessity for the human spirit to flourish. Places where humans tread lightly on the ground are not just our greatest reserves of wild creatures, but also provide immeasurable benefits to the humans who visit them. By 2014, the acceptance that some of Scotland's wilder areas should (in principle) broadly be left to their own devices for the benefit of humans and the environment alike had become government policy, and Scottish Natural Heritage designated forty-two Wild Land Areas, of which the Cairngorms was the largest. These areas came with a limited set of planning restrictions, designed to give them a loose protection from development.

Democratic, ecologically minded, forward thinking, the Unna Principles show us how a number of conservationists perceived the mountains at the time. But they were not perfect. For one thing, the mountains of Scotland are not an inviolate wilderness, left in its natural state. While humans generally tread lightly in our mountain areas, their wildlife and ecology have been deeply damaged by centuries of use. As we shall see, high up at the altitudinal edge of where the trees can survive, is a habitat that has been so obliterated by humans in Britain that most people fail to notice that it was ever there at all. With the loss of this 'montane scrub zone' have gone many ways of life, common in other mountain regions. Unna's principles underappreciate the scale of damage that humans have already wrought on Scotland's mountains, and so make no allowance for restoration work, which goes beyond keeping them conserved 'in their existing state'. Unna suggested that the mountains should be open to all, but in practice he meant only those who possess the ability to do so without the assistance of hostels or even signposts. Democratic, then, but only up to a point.

By 2002, the Trust was finding the Unna Principles were becoming cumbersome in a cultural environment which has come a long way since the 1930s, and so it introduced a new, pared-down Wild Land Policy with a new definition of wild land. This definition provides the guiding principle to the management and protection of the mountains under Trust protection: 'Wild land in Scotland is relatively remote and inaccessible, not noticeably affected by contemporary human activity, and offers high-quality opportunities to escape from the pressures of everyday living and find physical and spiritual refreshment . . . The primary purpose will be to identify, protect and enhance the "core wild land" areas of Scotland.'[7]

The phrase 'not *noticeably* affected' is telling. Here, then, is a less dogmatic, more outcome-led philosophy of mountain management than that proposed by Percy Unna. The Trust's definition accepts the philosophical tautology at the heart of human management of 'wild land' as a means to an end, for the ultimate benefit of the mountains themselves. Our wild land may not be truly wild, but it is the best we have, and is more than worth protecting from the worst in ourselves, for the benefit of all of us.

The first challenge for conservationists is to open up the mountains to as many people as possible, without allowing people to destroy the things which make the place so special. The second challenge is to undo the damage that has already been wreaked upon our mountain landscapes. But an even greater challenge threatens our mountain areas. The spectre of climate change looms large over anyone who loves the high hills, and the work of anyone who would seek to protect them. The tops are home to some of the UK's rarest and most specialised subarctic and arctic creatures, specially adapted to this hostile environment. With warming temperatures, the high tops are already feeling the heat.

In this section, we'll be freeing one of Britain's rarest plants from its cliff-edge prisons. We'll be looking at long-lying snow patches, and researching their worrying new inclination to melt. We'll take a closer look at the footpaths we use to access the mountains themselves and study the dotterel, continuing the work of one of the twentieth century's greatest ornithologists. We'll be playing with willows and birch, trying to support the last remaining specimens of once-common species, and return a lost habitat and its myriad ways of life to the Scottish mountains. In the words of John Muir, pioneering Scottish environmentalist,

> Thousands of tired, nerve-shaken, over-civilized people are beginning to find out that going to the mountain is going home; that wildness is necessity; that mountain parks and reservations are useful not only as fountains of timber and irrigating rivers, but as fountains of life.[8]

We have a complicated, contradictory, constantly evolving relationship with the mountains, but that's on us. That the mountains are understood almost universally as a panacea for the nerve-shaken and over-civilised says more about the unnatural environments of civilisation than it does about any inherent qualities of wildness to be found in our hills. That we should feel more at home in the ravages of a Cairngorms winter than in our own towns and cities is a savage indictment of how out of touch we have become with the things that truly enrich our lives.

12

Alpine Sow-thistle

If the mountains are a place of remoteness and danger, then they are also a source of refuge, a safe haven away from the wilder ravages of the modern, human world. And they are never more so than for the showy, extremely rare, alpine sow-thistle, whose entire UK population can be found on just four mountain ledges dotted across the Cairngorms National Park.

It's a Monday morning, February 2019, and I have found myself sitting in the atrium of the science block of the extremely venerable Royal Botanic Garden Edinburgh (RBGE). It's a bright, sunny winter day and, despite the seasonal lack of flowers, the gardens are looking stunning. The science block is the sort of place in which important-looking people bustle about doing important-looking things, muttering in Latin. There is a disarming amount of history and science and scientific history posted about the important-looking polished stone walls, and beautifully produced pamphlets for visitors to read in a waiting area. After a decade of austerity in Britain, all of the environmental offices that I know of are looking a little neglected, but somehow the RBGE didn't get that memo. I always feel slightly out of place in a city, but in these learned halls, dressed in dirty jeans and a fleece, I feel more than a little out of my depth. The science block of the RBGE is a place that sufferers of imposter syndrome would do well to steer clear of.

When, the previous Thursday, I found that I would possibly be in Edinburgh, I rattled off a nonchalant email to RBGE's Dr Aline Finger, who works with alpine sow-thistle and a couple of other species that happen to also be the focus of special attention at Mar Lodge. I said that I'd like to write some stuff about alpine sow-thistle and the Mar Lodge experiment, and would it be possible to have a guided tour on the Monday morning, please? And then, thinking this far too little notice, I thought nothing more about it.

On top of a godforsaken hill in the Pentlands, fast in the teeth of a blizzard late on Sunday afternoon, I received a text message saying that yes, that would be fine, and come down whenever. Somewhat taken aback by both the kindness and the speed of the response, I spent the night before the little tour dreaming that I was trying to get to the Gardens and kept getting on public transport which I thought would take me there but inexplicably moved me further away. Also in the dream I was wearing flip flops which kept falling off whenever I ran for a bus.

Fortunately, my guide for the day is Gavin Powell, who has been working 'front-of-house' showing folks around the Gardens for five years and has been volunteering with the sow-thistle project, so is very good at dealing with flustered visitors. Right now he's on secondment to the project and is helping out with the day-to-day tasks that come with conservation horticulture. He's exactly what you would hope for from a guide: passionate, clever, with just a hint of diffidence that humanises the whole lofty edifice of the place. In fact, all of the workers at RBGE seem to possess the same charming mixture of extreme competence and importance with diffidence and humility. They seem to feel the privilege of not only getting to work with the plants that they love, but also the privilege of public service, which is nice.

Gavin walks us around the block, away from the extremely smart public garden and visitor centre, into the guts of the operation, a series of plantings and greenhouses in the nursery. I feel a bit more at home here. As we're chatting away, talking shop in a large greenhouse, I suddenly realise that we're surrounded by some of the rarest plants in Scotland. Here is a pot of twinflower, a rare pinewood specialist that we also happen to be working with at Mar Lodge (work which would be worth a chapter in itself). Next, a row of globeflower plants; not particularly rare, and actually becoming reasonably common at Mar Lodge in recent years, but lovely all the same. Here's round-leaved wintergreen, another super-rarity, with weird spikes of flying-saucer flowers sitting over leathery leaves. Potted in with sapling rowan trees is small cow-wheat, a hemi-parasite that does well from having a tree to steal nutrients from. It has been reduced to a handful of outlying spots around the Highlands, including Corrour and Glen Strathfarrar. There is a small population at Mar Lodge, so small indeed that I've never quite managed to find it. Here are the montane willows, which we'll find out more about in a few chapters' time. These are Gavin's favourite: 'Each tree has such individuality, the way it tenaciously clings to cliff edges and scree.' Next, we walk past a small, unassuming group of saplings which, if numbers are to go by, are among the most endangered plants in the world: the Arran whitebeams, which are only found in a pocket of Trust-owned land on Arran. These species are the results of a complex series of hybridisation events between closely related trees of the *Sorbus* genus, and subsequent population isolation through their remote mountain location. One of these species, the Catacol whitebeam, is currently known from just one wild tree. For the Scottish botanist, visiting this greenhouse is a thrilling, though somewhat complicated experience, somewhere between seeing a botanical supergroup playing at a festival and visiting a Madame Tussaud's exhibition where the waxworks are not only alive but also cloned from the celebrities that they mimic.

Then there's the alpine sow-thistle. Pots and pots of it, each one individually marked with its own label and barcode.

Alpine sow-thistle is a peculiar mountain plant in that it is enormous. When we think of mountain plants, we tend to think of stunted trees, low-growing, wind-clipped heather, saxifrages with flowers not more than a centimetre across. And yet as long as it has lots of water and something resembling shelter and not too much direct sunlight, alpine sow-thistle grows large, luscious, dandelion-on-steroids-style leaves, and sends out a huge flowering spike, which can reach as tall as two metres, capped by a raggedy blue flower. In Norway, it can be found at sea-level, while in the Alps it is common enough in the high-altitude meadows. While probably never common in Scotland, at least in the last couple of centuries, it used to grow on valley floors, along the edges of burns and possibly in high-altitude birch woodland. In the twentieth century it suffered a major range contraction, away from the valley floors, and up into its current mountain fortresses. The reason? You've guessed it. Sheep, goats and, latterly, deer. It is, as Cairngorms National Park ecological adviser Dr David Hetherington calls it, a 'canary in the mineshaft of grazing pressure.'[1]

'It's extremely palatable,' says Gavin, 'so much so that it has another name: mountain deer's lettuce.' Given the choice between heather and a sow-thistle salad, I know which one I would take, though Gavin assures me that the sow-thistle is rather bitter.

It was only in the high mountains that sow-thistle managed to cling on, and now its future lies literally on a cliff edge. Its final ledge homes are between 700 metres and an improbably high 1,050 metres above sea level, beyond the range of almost all of Scotland's plants. In fact, beyond botany nerds the only people to have any real knowledge of the plant in Scotland are climbers. One population, which sits just off a well-known climbing route, is known as 'the Potato Patch', which says as much about the botanical knowledge of your average climber as it does the parlous nature of alpine sow-thistle's fingerhold in Scotland. With extreme isolation comes a degree of safety from nibbling critters, but also a trap. Stuck on its ledges, the plants are isolated, so they cannot reproduce successfully with other populations. 'The team have done a lot of genetic tests on the plants. All of them are very closely related. One or two of the ledges can produce fertile seed in the wild, but not very often at all. And the stuff it creates is poor quality; it really struggles to germinate.'

The four cliff edges may provide a refuge, but they are, ultimately, also cliff edges, which are dangerous places. In 2016, a landslide ripped through the Potato Patch, which, as Gavin says, 'wasn't entirely helpful'. He goes on, 'The work of people like Dr Chris Ellis, who has headed up the team from the beginning of the project, tells us that without human help, the Scottish population has very poor prospects for the future.' Scientists are not prone to hyperbole, so what this means is, unless we do something, it is doomed.

The prospect of us doing something has been lurking in the background since the late 1990s, when surveys and genetic studies were first done on the ledge populations. Since then, RBGE has put together a team to safeguard Scotland's alpine sow-thistle populations for the future, which has become well supported by any number of external partners.

Every few years botanists from RBGE, assisted by professional climbers, have abseiled down to the ledge populations to count plants and measure their spread. Stage one of the sow-thistle conservation plan was to create an ex-situ population of the plant away from its fingerhold locations. Small samples were taken from each site, just enough to propagate. This was to be held in reserve as a 'bank', just in case the worst happens with the wild populations. Stage two of the project was to divide and multiply the stock for trial translocations. Stage three involved cross-pollinating stock, to increase genetic diversity and create more material.

Gavin is keen to show me as much of the plants are possible, so we retire to a workshop with a potted plant. 'The plant you see above ground is really just the tip of the iceberg,' he explains, while delicately prising apart a mass of roots and rhizomes. 'The real magic goes on below ground.' He washes the soil off the rhizome, and we come across a ball of life 'a bit like a kraken'. There is something almost alarming about how vivacious the thing seems. It is this powerhouse of life and energy which is the plant's secret weapon: with a deep root network, it has reserves of energy and nutrients which it can draw on to produce huge growth very quickly. This is good in an environment with short summers and a profusion of grazing animals.

By carefully prising apart and separating the rhizomes, the team can very quickly grow on new stock. This is a useful part of the project, but it is not perfect. Each rhizome is a clone of the same plant, so genetic diversity is not improved through this process. To secure the plant's long-term future, new genetic diversity has to be introduced. Gavin shows me a tray of tiny, germinating plants, derived from Scottish seed created by cross-pollinating material from two different ledge populations. They are second-generation plants, which is a milestone for the team and a major step forward for the genetic diversity of Scottish alpine sow-thistle. But this profusion of life comes with its own challenges. The most pressing day-to-day concerns for the team are now logistical. Each individual plant is labelled and barcoded, so that the exact genetic make-up of each plant is known and easily traced back. It is time-consuming and complicated, but vital to ensure the long-term viability and variability of the stock of material.

This is particularly important because there is not just Scottish stock here. 'We have some Norwegian plants here too. This is for comparison purposes – we're keeping the Scottish and Norwegian plants separate. This is entirely subjective, but to me the seeds seem bigger and fatter than the Scottish ones, and they do seem to germinate more successfully than the Scottish stuff. But it seems that cross-breeding the different Scottish populations is helping.'

Why not just introduce stuff from Norway? I ask, somewhat trollishly.

'We're talking about pretty intensive conservation in really sensitive species and locations, so there's an important ethical element to our work that we have to consider. We're protecting the Scottish population, not the Norwegian population, which is actually pretty stable.'

It's a tricky balancing act: what species belong in a landscape? How much intervention is too much intervention? When do we cross over from conserving our wild areas to just gardening them? How much of our understanding of environmental conservation is based on the cultural conventions of national borders?

Cliffs are all well and good, but for the plant to survive in the long term, it will need to once again thrive in its 'original' habitats – upland meadows and damp woodland, among the dappled shade of the birches. This is stage three, and it is now in full swing. With a bank of genetically enhanced stock at their disposal, the team set about finding places to put it. At this point the experiment took a new direction: where would be suitable? Should the plant remain restricted to cliff edges? The end goal is to produce healthy, viable and self-sustaining populations into the future. This means that alpine sow-thistle will have to come down from its refuges. Three sites were chosen across the Cairngorms in which to plant out stock.[2] Mar Lodge had habitat which seemed to fit the bill for 'replica Norway' habitat: wet, upland birch and pine woodland. It sits within the plants' historic Scottish range: it is thought that the plant was lost from Mar Lodge fairly recently, but no one noticed. It also has low levels of browsing over a large area and outside of stock fences. This is a very precious commodity in Scotland.

Of course, things don't always go entirely to plan. In autumn 2017, 160 plants were introduced by RBGE staff to two secret locations at Mar Lodge. We kept a close eye on the plants. We watched as they sprouted their first salady leaves the following spring. But things then took a turn. By May, we were expecting the worst. By the end of the growing season, it was apparent that the planting sites were a scene of carnage. All of the plants had been eaten by a mystery herbivore. Fortunately, much more successful results came from another experimental planting site within the National Park, higher up at 800 metres altitude, where they managed to get 80 per cent survival rates, and even had plants flowering. Nevertheless, by the time I visit Gavin in Edinburgh I was feeling a little depressed, and slightly ashamed that Mar Lodge hadn't seemed to have been up to spec. 'It's still early days,' he says tactfully.

In science, a failure is not a failure unless you fail to learn from it. In the 2019 growing season we up the ante with our monitoring efforts. I say we, but actually the monitoring is done by a team of local volunteers. We install a series of fencing controls around various plants. Some plants are deer-proofed, others have a fine, hare and vole-proof mesh installed around them, and some are left without any protection at all. We then install camera traps to see what will turn up and eat them.

The volunteers make regular trips to the sites to see how the plants are getting on. In early April I take a stroll up to one of the sites. It's tucked away along a rather special burn, in amongst pine, birch and aspen. The mesh cages look incongruous against the clear water and the birchbuds. For now, there's not much to see. I notice that the basal rosettes of the sow-thistle are further on than other palatable species nearby and are therefore highly conspicuous to anything that may wander by and feel peckish. I wave at the camera, and idly wonder whether a slug will crawl through the mesh, have a go at the lettuce, and grow so large that it cannot escape again. I try to imagine what the burn will look like bedecked with head-high blue flowers, a riot of colour. I struggle to comprehend the luxuriant growth – the Highlands have for centuries been a place of slow growth, constantly nibbled away. The image somehow eludes me, for now at least.

June, and things are hotting up. The cameras are providing brilliant, if anticlimactic, data: mountain hares and black grouse set off the traps, as do a few walkers' errant dogs, but nothing appears to be eating the plants. Except slugs. Giant, thick, glossy black slugs. Somewhat contrary to the methodology of the experiment, the volunteers flick them off whenever they see them, and we pretend not to see them doing it. As the summer rolls on it becomes clear that we have found our mystery herbivore. Not voles, mountain hares, or deer. Nothing so lofty. Our mystery muncher is, of course, what any gardener in Scotland who has tried to grow lettuce could have told us it would be.

I find myself feeling a loathing towards slugs such as I have rarely felt towards any creature. I now understand the instinct of the gardener to destroy the oozing, slimy sausages. Before, I had never much noticed them, nor even thought to think about them. Now, as I watch one chowing down on the sow-thistle, I feel a churning in my gut. I tell myself, over and over, that it is just a part of nature – a fact of life, a beautiful part of life's intricate tapestry. The ecologist in me, the supposedly rational, distant observer of natural phenomena, comes up against a profound, irrational hatred of a newfound foe. I fight the cognitive dissonance that so muddies environmental conservation efforts – the eternal struggle between head and heart. The black slug laughs in the face of scientific disinterest, and I prise them off the plants with glee.

In July, a plant sends up a flowering spike. I check it as often as I can, and finally we are rewarded – the first flowering alpine sow-thistle at Mar Lodge in living memory. It is a single flower – a symbolic success, but the project has a long way to go. We are understanding the needs of the plant better, and where it likes to grow, and indeed how best to plant it. We now think that 'mob-planting' is the way forward – planting as much stock as possible in a reasonably small area, so as to overload the local herbivore populations. Alpine sow-thistle, it seems, is an all-or-nothing plant, which does well in large patches but struggles to survive in small pockets. The plant's strategy may be to overwhelm other plant species by growing quickly over a large area, accepting that a high proportion of its growth may be eaten by herbivores. We think that large losses are acceptable, as the plants are long-lived. They don't have to be successful every year. Losses to slugs may, perhaps, just have to be factored in.

Now, as I walk through the birchwoods, I see the landscape in a new light. I find myself picking out spots that would be suitable for the plant. I scan the high corries for inaccessible ledges which might just harbour another elusive refuge patch, waiting to be unlocked from its airy prison. Alpine sow-thistle is slowly returning not just to the Scottish landscape, but also to the Scottish imagination.

Things, then, may be looking up. Gavin tells me that another ledge with a few extra plants had been found nearby one of the known populations. It was noticed by Mike Smedley, NatureScot's alpine sow-thistle guru, while taking his dog for a walk in the mountains. Meanwhile, these experiments are telling us not only more about alpine sow-thistle, but also ecological interactions across a range of montane habitats. They are showing the team at RBGE what is and isn't possible for the species, helping us to reimagine what plants used to flourish in our mountains, why they were lost, and what might flourish again in the future.

But there is another reason for the work, one beyond aesthetics and the ethics of reintroduction and translocation. Looking at the willows in the Botanic Garden, Gavin spoke of his love for the Scottish mountains. And then he finished with a throwaway comment. 'Scottish mountains are a hugely special place, partly I suppose because they're much lower than the mountains of Scandinavia and the Alps. This means that they are more susceptible to climate change. Our work really is as cutting edge as you can get. What happens here, and how well we work to mitigate it, may well go on to inform what happens across Europe in the coming decades.'

Good news, maybe, for now. But the dark clouds of climate change still loom on the horizon. Let's look at some snow.

13

Sphinx

Mid July, and as is often the way at this time of year, the weather is somewhere between inclement and filthy. Six of us, five botanists and myself (a mediocre botanist at best) are sitting outside the Garbh Coire emergency shelter, grazing on oatcakes, sausage and cheese. Ian Francis, vice-county recorder, has corralled us here to roughly the most remote spot in Aberdeenshire in the name of science.

The Garbh Coire emergency shelter, a tiny construction of steel and wood, sealed off and camouflaged from the world by some serious drystane walling, sits in the middle of a bewildering array of corries. We are surrounded by three great lumps, Cairn Toul, Sgor an Lochain Uaine and Braeriach. This is a glacial landscape, simultaneously young and ancient, a properly wild place. Ten thousand years ago, several glaciers merged together here and flowed south, following the line of what is now the River Dee, taking half of Bod an Deamhain, the thousand-metre-high Devil's Point, with it.

The wildness here is verging on oppressive. The place feels raw, edgy. We look over to the western slopes of Ben Macdui. We're sitting around 700 metres above sea level, but Macdui keeps pushing up, just one huge boulder field. The cloud layer today is hundreds of metres below the summit.

These corries are where the fun plants are, so Ian's corralling has less to do with getting us to this hostile spot in the first place, and more to do with stopping people wandering off willy-nilly in quest of botanical scarcities. This is one of the largest places in Scotland where rare and exciting arctic and alpine plants, specially adapted for cold and snow, can survive. Lower down the glen we have already found some good stuff. We have passed bushes of *Salix phylicifolia*, tea-leaved willow, looking well and expanding outwards, thanks to the reduced browsing pressure from deer. Others have picked out sheathed sedge, a tricky species to spot and identify, and northern rock-cress, a rare gravel and shingle specialist. We find two candidates for *Betula x intermedia*, a hybrid between dwarf birch and downy birch, recorded only a hundred or so times in Scotland. Great sundew, the showy, spectacular carnivorous plant, turns up with a regularity that surprises everyone. Its menacing leaves, covered in red filaments capped with a tacky glob of gluey secretions are a real crowd-pleaser, and it is a rare enough plant this far east. Now, we sit, enjoying a brief lull in the rain, mustering forces for the final push into the really good bit. A roebuck and doe idly waltz through a patch of globeflower a couple hundred metres away. Here, as we sit eating lunch, the plants are teasing us. It is exactly the excitement of a kid going to Disneyland, waiting in the queue, knowing that inside lurks paradise, but that it is, temporarily at least, out of reach. It's too much for Dan Watson, ecologist at Ben Lawers, and a particularly skilled upland botanist. He breaks rank, bounding up the steep incline, scrambling around a precipitous, slabby waterfall to have a look at a patch of willows.

A switch is flicked, and now everyone's off, Ian rushing behind, flapping recording papers and grid references. It's gold fever, but for plants. Beyond the shelter things get interesting, quickly. We find large patches of alpine meadow-rue, globeflower, alpine saw-wort, dwarf cornel, Scottish asphodel, three-leaved rush. Good plants, but 'gateway' plants, suggestive of rarer treats to come. *Sibbaldia procumbens*, least cinquefoil, but known by all just as *Sibbaldia*, puts in the first of several appearances. That's more like it.

Nice as it is to go off on botanical adventures into the hills, there is a purpose to all this. Every twenty years the Botanical Society for Britain and Ireland (BSBI) publishes an atlas of botanical records. Ian Francis is in charge of generating plant records from South

Aberdeenshire, and he's mortified to discover that there are very few records from the west of his beat. In fact, in the last twenty years no one has bothered to report any plant records from about sixty-five square kilometres of Mar Lodge Estate land. The fact that this is the most remote part of Aberdeenshire is immaterial, as is the fact that such reports, vitally important though they may be, are entirely dependent on the efforts of volunteers – what's the point of having an atlas if it doesn't have 100 per cent coverage?

We push on, up into the corrie of the Falls of Dee, just below the source of the river itself. This is a curiously hospitable corrie, despite the weather. Snow buntings call and chirrup, a ring ouzel passes through. There is green-ness in this corrie, which hints at a base-richness in the nearly non-existent soils, which in turn hints at botanical rarities. It is remarkably verdant for somewhere that is quite a bit higher than the summits of most Munros. And now we are indeed into the rarer stuff. Alpine willow-herb, arctic and alpine mouse-ear, alpine lady fern.

The graminoids, the grasses, rushes and sedges, are getting everyone particularly excited. Grasses are remarkably beautiful plants, once you get to know them a bit. There's *Phleum alpinum*, alpine cat's tail, fewer than 300 records in Britain. *Poa alpina, Carex x decolorans, Carex lachenalii, Carex capillaris, Juncus triglumis*. The next day Dan Watson will outdo himself and find *Carex x biharica*, fewer than ten records in Scotland, while scrambling up a ridiculously perilous-looking gulley (at least to a non-climber) to find a cudweed. Later in the year he will go on to find another sedge hybrid, even rarer, at his usual stomping ground, the obscenely botanically rich Ben Lawers.[1]

Eventually, inevitably, we wend our way into the Garbh Choire Mòr, the big rough corrie. It's another beast entirely. There are four of us now (it's getting on, and two of the party have retreated to the relative comfort of the tents), scouring the corrie, testing nerves in the gullies, where boulders are loose and moss-covered and the ground shifts under your feet. There is a wobbly feeling to the ground here, and it isn't pleasant. This is an oppressive place, shapeshifting, mercurial. The wildness is energising. We push higher. There are fewer plants; this feels inhospitable, rawer and edgier than anywhere else we've looked. The cliffs are dark and wet, the sun rarely reaches into the gloom. But we find a few nice things. I nearly tread on the only *Alepocurus magellanicus* we come across – a properly rare grass. These plants are the hardiest in Britain, but they are also among the most endangered; they are wild and fragile.

There are echoes of the Victorian collectors in our endeavour, those hardy individuals who both massively increased our knowledge of the plants we share our land with, but whose avarice for rare specimens also brought many of them to the brink of extinction in the process. We are only a few sets of tweeds and a field glass away from their work, in all honesty. But rather than extracting and taking home specimens for personal collections, the prize here is simply to see, photograph and tick off the plants. It is very cold and very wet. Paper goes mushy, plant books remain firmly ensconced in plastic bags, buried deep in increasingly saturated rucksacks. Gloves go on. Tech and kit may change, but a spot of botanising in Garbh Choire Mòr remains as difficult an undertaking as it was when people first came here and felt like it might be worth rummaging around for some rare plants. It is made into a prize worth having by the difficulty and discomfort of the endeavour, and the scientific exigency of the task. Botanising in this way feels like a remarkably pure form of science.

We turn to leave. It's getting on and we've a couple of hours to walk back to our wet, midgy camp at Corrour bothy. The cloud lifts to 1,200 metres. Leering above us, drifting in and out of the mizzle, sits the Sphinx, an ominous, dirty white mass of snow and ice, as it would have sat for those first botanists a century and more ago. The plants we've been searching for, the arctic-alpines, are the echoes of the Sphinx sitting above them.

People have been in love with Scotland's late-lying snow patches for centuries. They are the stuff of legend. The Gaels named corries after them, like Coire an t-Sneachda on Cairn Gorm. Ancient records of snow come from locals, drawn to record both the quotidian and unusual aspects of their lives. Snow tempts the curiosity of the tourist. Snow signifies wildness, and so to see summer snows confers an air of respectability on the travels of the would-be adventurer.

Tracking the fortunes of Scotland's snow affords people a visceral sense of the state of the land in which they live. These patches are a living, dynamic element of our landscape, linking us to our ancestors, and the Sphinx is a late-lying snow patch with a hint of notoriety about it. Named after the climbing route that it sits below, it is the longest-lasting snow patch in Scotland. It is unusual for it to melt. Snow lies here all year round. This is the reason that the corries harbour all those rare plants. Late snow equals arctic-alpine plants, mosses, liverworts and lichens. Mountain hares and ptarmigan use snow patches for camouflage and insulation and protection from storms. Dotterels and snow buntings pick insects from the edge of snow patches. Palatable plants like willows do well from having it cover them over for the winter, protecting them from herbivores. The benefits that snow bestows to these ridiculously hardy creatures are inverted in less hardy creatures. By creating ecological niches, snow confers an ecological advantage on some species over others. In Britain, because of the relative paucity of snow, these creatures are all rare. And now the snows are disappearing.

It is here, among these records, that we meet Adam Watson again, his long shadow like the Grey Man of Macdui, drifting in and out of the mizzle. For seventy years (seventy!) he studied the Cairngorms' snow patches. It is largely thanks to him that Scotland's snow patches are among the best documented in the world. He followed the Sphinx for his whole life, and reading his work, you get the impression that it was one of his favourites. In 2010 Watson and Iain Cameron, *the* Scottish summer snow expert and Watson's snow acolyte, wrote up a book, *Cool Britannia*, of all the historic records of Scottish summer snow that they could find, going right back to the sixteenth century.[2] So remote is the Garbh Coire that the first reference Cameron and Watson could find for it in the literature of snow comes from the early nineteenth century. This is unusual, because records of Scottish hill snow are reasonably common from the seventeenth century onwards. This first record gets the somewhat foreboding nature of the Garbh Choire Mòr pretty well: 'Even in the sunniest weather it is black as midnight, but in a few inequalities on its smooth surface, the snow lies perpetually.' Half a century later, Professor William MacGillivray, botanist and naturalist, noted that the Garbh Coire was 'an immense unmelted mass of snow'. On 27 August 1918, D. P. Dansey wrote that 'it is the "Eternal Snow Corrie" of Braeriach, and the snow at its head has never been known to disappear'. Another famous Cairngorms naturalist, Seton Gordon, indicated that he had never seen the nearby Pinnacles site without snow up to the mid 1920s.

Between them, Watson and Cameron amassed enough data to strongly suggest that the Sphinx snow patch was a permanent fixture in the Cairngorms landscape from at least 1700, and probably long before that. The snow at the bottom of the Sphinx patch, buried by successive years of snowfall, would be years old, decades even.

Or at least it was. The perennial snows of the Sphinx are melting into history. In 2019, when we look up from our plant specimens and see it, looming up out of the cloud, the snow is less than a year old. In Watson's lifetime, the Sphinx patch melted seven times. Three of those times were in the twenty-first century. Two of those times were in 2017 and 2018; the first time in recorded history that the patch melted in two consecutive years. Climate change is hitting Britain, and Scotland's summer snows will be one of the first casualties. In 2019, the Cairngorms National Park Authority published a study into the effect of climate change on the National Park's snow. It found that there is likely to be a substantial decline in the number of days of snow cover from 2030, potentially leading to biodiversity loss, an impact on local water supplies and potential for increased flooding due to rapid snow melt.[3]

Since Watson's death in 2019, Cameron has taken on Watson's mantle as snow-watcher general. Every year he organises a mass snow watch across Scotland's hill ranges. In the first weekend in July, watchers are sent up across the Highlands to record any snow patches that they can find. In 2019, the weekend coincides with a very short but excellent weather window. It also coincides with the passing of the Sentinel satellite, a nice piece of kit that photographs the world across various light spectra, and whose images just happen to show up snow patches extremely well. Never one to miss an excuse to head into the hills, I offer to walk up Macdui and back through Coire Etchachan.

It is one of those days that people who go into ecology expect every day to be like. A snow bunting, one of Britain's rarest breeding birds, sits on Macdui's summit cairn, and I try to feed it some sandwich crumbs. I tell a couple of walkers about it: I'm always keen to pretend I'm a nature guide, and the pair don't seem to mind my intruding on their lunch too much. Reindeer from the semi-wild herd cool off on the big snow patch between Ben Macdui and Cairn Gorm, and walkers crowd around them to take pictures. The patch covers several hundred metres, enough to suggest it would never melt.[*] I take high-resolution pictures of the Garbh Coire complex and will later stitch them together to show the extent of the snow cover. Not quite Sentinel-level tech, but hopefully useful nevertheless. I look into the Garbh Coire from across the Lairig Ghru and take in the Sphinx. It seems innocuous enough from there. This day is a day like Queen Victoria had on 24 August 1860, when she rode a pony from Mar Lodge to the summit: 'The view at the top & from the cairn wonderfully fine & extensive, ranges & ranges of mountains, rising one behind the other like the waves of the sea & all so blue.'[4] But then Macdui has its own laws. One moment it is a still haven, the next, as Nan Shepherd says, it is a roaring scourge.

Walking down to the crystal-cut blue of Loch Etchachan, I stop to inspect a snow patch in closer detail. Summer snow is a curious thing, far removed from the blizzards of February. It is covered in oblations and undulations like the surface of the sea, waves of air

---

[*] Spoiler alert: it did, quickly.

frozen in time that follow the prevailing wind. The top surface is something like slush puppy, while underneath the consistency is closer to ice. The patches bleed water in the hot summer sun. Deep channels open up underneath patches, along waterways, allowing the foolhardy to go 'caving' deep beneath snow patches. The edges are dirty brown, somewhat detracting from the otherwise pure, clean, blue-white of the patch. I crunch along its edge, feeling guilty that doing so will speed up its contraction, opening up new surfaces to the sun, wind and rain, but nevertheless enjoying the contrived feeling of the tourist-adventurer, traversing summer snow in a remote land. Look at the map, plot all the points, and you'll see that snow patches in the Cairngorms tend to face north-east. They tend to sit just under the rim of a face. Here is why the Sphinx is such a long-lived snow patch. It is high, so it is cold. It is sheltered from the wind and rain from the prevailing south-west. But conversely, snow is blown into the corrie by winter storms, so the conditions which provide shelter and shade in summer cause it to be deposited there in winter.[5]

A couple of days later Iain Cameron gets back to us with the results. They are better than expected, given the shockingly low amount of snowfall in the previous winter. This is thanks to the late, unseasonal snow in May and comparatively cool temperatures since then (the same snow that messed up my spring monitoring of the Mar Lodge hen harriers). But it will be touch-and-go whether the Sphinx, and therefore any Scottish snow, survives the year. This is the nature of the puzzle of the Sphinx. It is at once fragile and ferocious, wild and threatened.

Late October 2019, and the snows are beginning to fall again on the high hills. I've been thinking about the snow patches a lot this year. In my head there is a great deal riding on the survival of a small speck of ice or its disappearance. It is an act of hope – perhaps we haven't messed up our climate as much as we think, perhaps nature will fix itself. Because if the snow still sits in Garbh Coire, perhaps we have time to save things, perhaps it's not yet too late, hope against hope. The Sphinx is a bellwether. Its survival into the new winter signifies continuity, the continuing wildness of our wildest corrie. Its loss suggests the opposite.

There's a growing interest in Scotland's late-lying snow patches. There are probably a few reasons for this. For one thing, they are much more accessible than they used to be. Thanks to an active social media presence, and more people heading to more remote places than before, Scottish summer snow is almost becoming mainstream. Iain Cameron now regularly appears in national newspapers, on TV and radio.

Scotland's summer snow has always been interesting because there's not much of it – it is a quirk of nature. In a national context it also signifies the uniqueness of the Highlands. But nowadays, following the story of Scotland's summer snow holds the same fascination as watching a car crash in slow motion. Watching the snow patches disappear earlier every year is an exercise mediated by what the anthropologist Andrew Whitehouse calls the 'anxious semiotics of the Anthropocene'[6] – the niggling feeling that things aren't right, the intrusion of human activity into the wild. These days, I watch the weather forecast closely, looking for how much snow is likely to fall into Braeriach's corries, how much is likely to melt. It is useless for me to do so, but it gives me a false, comforting sense of power. Like too much in environmental conservation, it is simultaneously an act of mourning and a forlorn hope for the future.

I've followed the weather extremely closely for the last couple of months. It has been windy, warm and rainy – not good for snow survival. The other day, snow fell at the Lodge. I picked some up and thumbed a snowball together, an act of reflection and, possibly, penitence. This year there are just two surviving patches in Scotland, the Sphinx and the nearby Pinnacles patch. But these are not the great walls of snow and ice of previous years. They have been reduced to tiny fragments, just a few metres each. It's a damned close-run thing.

There has recently been more activity online. Three tiny dumps of snow have fallen on the Cairngorms, interspersed with good(ish) walking conditions, so pictures have been trickling in of Garbh

Coire. They're inconclusive. There is snow in the Sphinx patch, but it is fresh, and obscuring what would have been a tiny patch, maybe a metre in size, nothing more.

I refresh the Snow Patches in Scotland Facebook page, nothing. Iain Cameron was planning to inspect the snow patch this weekend. I expect I'm not the only one casually-not-casually refreshing the page.

Finally, a notice is posted to Facebook. It reads as follows:

> An atrocious day's weather today on Braeriach. Incessant freezing rain and snow, with very poor visibility.
>
> The trip was over two days, as Alistair Todd and I had decided to overnight at the Garbh Coire refuge, for a bit of craic. Alas Al had to cancel due to ill health, but the seed had been planted in my head.
>
> On the walk in up the Lairig Ghru I met a couple who said they'd met four Czech hikers who were bound for the refuge. Not wanting to sleep outside I decided to head instead for Corrour bothy. A longer walk in but with the prospect of a fire and more room was too tempting to pass up. As luck would have it Neil Reid had a good blaze going when I arrived, and a few drams were taken.
>
> Anyway, to business . . .
>
> I left the bothy this morning in reasonable weather, but it quickly degenerated. The walk in into GCM was laborious, with the weather worsening by the hour. Strong winds and heavy snow falls made it really hard work. At one point I thought about turning back because the conditions were so grim. A couple of clearer spells encouraged me to carry on. What did I find? Did the old snow survive? Yes! Sphinx has made it but Pinnacles hasn't.[7]

It has survived, for another year. The globe is still working, for now.[8]

14

Footpath

> If the Scottish people yield up to his Grace their right of way through Glen Tilt, they will richly deserve to be shut out of their country altogether.
>
> Hugh Miller, 1847

Those Victorian botanists who forged the way for our own small endeavours below the Sphinx were quite literally trailblazers in the Scottish mountains. They worked without the benefit of the road and footpath system that we take for granted today, gaining access to the mountains only by way of horse and foot. They were walkers in ways that few of us would recognise today. The Victorians' prowess at the plod of walking still remains with us through William W. Naismith's rule. Naismith was a pioneering Scottish mountaineer who, in 1892, described off-hand 'a simple formula, that may be found useful in estimating what time men in fair condition should allow for easy expeditions, namely, an hour for every three miles on the map, with an additional hour for every 2,000 feet of ascent.'[1] The rule has been taken to heart, perhaps for its simplicity, and is now widely used for calculating journey times across the world. It's a useful rule of thumb, but as any but the fittest of hill walkers will have observed, it is a little on the fast side. Over the years multiple 'corrections' have been applied to it, in each case to slow it up. Finally, in 2019, the Ordnance Survey bit the bullet and decided to do away with it altogether, stating graciously that 'we felt our users – with their diverse abilities and ambitions – needed a more personalised recommendation, not one based on a single, fit, Victorian mountaineer'.[2]

Trailblazing, too, were the botanists in demanding access to our uplands. While most landowners were happy enough to oblige the occasional university professor and his entourage of students,[*] there were notable exceptions, and so Mar Lodge has its own place, if not directly in the long fight for access to our own countryside, then at least adjacent to it. In 1847, Professor John Balfour was leading one of his regular jaunts into the mountains with a party of students. All went as planned, as they crossed through Mar Lodge land to the march with Atholl Estate at the head of Glen Tilt. The party believed themselves to be on a right of way, but were met by a band of the Duke of Atholl's keepers, intent on keeping them out. After a brief altercation, a defeated Balfour and his students exited by way of a dyke.

It was to be a pyrrhic victory for the duke. Balfour, described by influential conservationist Derek Ratcliffe as 'something of a martinet in the best style of Victorian academics',[3] was a force to be reckoned with. He was a strong supporter of the newly formed Scottish Rights of Way and Access Society, who were spoiling for a fight with a landowning class that was getting ideas above its station. The lengthy lawsuit which followed vindicated Balfour's right of way through Glen Tilt. The charity, now known as Scotways, exists to this day.

Hell hath no fury like a botanist scorned.

Of course, the actions of a few botanists are only a small but illustrative part of the long battle to make our landscape accessible to its citizens. Footpaths are political statements: the radical history of Scotways, the Ramblers, the National Trust for Scotland and others is often forgotten, but at the core of these venerable organisations is an understanding that access to the land, and the protection of rights of way, is an integral part of a functioning society which is

---

[*] If not the general public.

truly 'at one' with its territory. A citizen is only truly such if they have access to their own country.

Centuries in the making, Scotland's upland footpath network spreads out across the hills, a tangled web linking citizens to their land. Footpaths are simultaneously manifestations of human mobility and also facilitators of it. In our footpaths lie our history and our culture. But just as footpaths are shaped by human conceptions of the landscape, they are equally shaped by the landscape over which they traverse. Footpaths seek out lines of least resistance. They curl around bogs and creep over boulder fields. A footpath through a wood will become overgrown and disappear remarkably quickly if no one keeps the wild away by walking it, while a wind- and water-scoured path up a Munro will collapse under the feet of its users unless it is maintained regularly. A footpath is only a footpath as long as there are people to walk it. As such, footpaths are the purest physical manifestation of our ancient links to the land. They remind us of our place in the landscape. Scotland's upland footpath network tames the landscape, but it enwildens its people. While a hill track, bulldozed through a landscape to make way for vehicles, is a brutish imposition on the landscape, an ancient footpath is a reminder of our place in it. Small wonder that one of the first jobs undertaken by the Trust at Mar Lodge, back in 1995, was to restore fifteen miles of vehicle tracks into footpaths.[4]

Footpaths are magic, in that they contract space. By making walking easier they bring points closer together in time for the walker. They are political statements – this land is ours to walk on or pass through. Some footpaths are ancient lines across the country, drovers' roads, trade routes like the Lairig Ghru. Others 'pop up' to suit the need of a new trend or industry. Some, like stalkers' paths, change over time, or disappear altogether, like the Duke's Path up Braeriach. What were once built as a means to an end for deer stalking, often improving more ancient thoroughfares, are now widely used by a community of tourists, wilderness seekers and Munro baggers. Small wonder, then, that footpaths have been at the heart of the conservation movement for as long as such a movement has existed. From the legal tussles of the Tilt Road and Jock's Road, through the famous 1932 mass trespass of Kinder Scout in the Peak District and right up to the adoption of the Outdoor Access Code and beyond, the right to roam our land on the footpaths used by our forebears is one of the oldest of the green movement's battles.

New footpaths spring up out of notoriety. A lack of footpaths will deter people from an area for a while, but if walkers are keen to visit an area then a desire line will quickly become a track, then a path, then an eroded path and finally a scar. Footpaths, then, are most often appreciated in their absence. It is a truism that the Grahams are the hardest of the walkers' hill sets to complete due to the lack of footpaths. It is all yomping through bogs and deep heather for those rarely visited peaks. The best footpaths are engineering marvels. Unobtrusive, subtle, hard-wearing, the best footpath is one that isn't noticed at all, a subtle ribbon through the landscape.

We have all felt the tension between sticking to the path and rambling off for adventure through the bog. Such is the paradox at the heart of wild land – we all want access to it, but too much access would destroy it. To complicate matters further, everyone has their own idea of how much access is the correct amount: as the well-worn saying goes, 'One person's wilderness is another's roadside picnic spot.'[5] The footpath sits somewhat uneasily in this ontological puzzle. While it facilitates access to wild land, thus reducing the wild land quality of the area, it also acts as its conservator. It does this by keeping people to one area – it is a ribbon of sacrificial land, highly developed, which allows the remainder of the land to stay undisturbed by feet and wheels and walking poles. There will always be arguments about the 'correct' amount of development along footpath networks, and indeed the impact of specific footpaths in specific places. But the best footpaths add to the wildness of the land, rather than taking away from it.

Waxing lyrical about the sublime qualities of footpaths is all very well, but footpaths are of the earth, as it were, and so there are practical and financial matters to be considered. The increasing popularity of walking in wild areas is, like all development in fragile ecosystems, both a blessing and a curse. Footpaths are worn down by, well, feet, and increasingly by mountain bikes. Overuse of footpaths can cause a real headache for cash-strapped land managers. As always, appreciation of wild land can become detrimental to the land itself.

So if you take away one thing from this book, let it be this: if you are using Scotland's upland footpaths, you should really be thinking about donating money towards their upkeep.

It is easy to take our access to the mountains for granted in Scotland. South of the border, the rules for ramblers are far more restrictive. Many countries demand fees for access to their wild land areas, like the US National Park system. Other places, like Sweden, provide excellent facilities in their wild areas, but only at a hefty price tag. Scotland's curious mixture of excellent outdoor access and extremely concentrated land ownership is pretty much unique in the world. This brings both positives and negatives for our mountains. Despite being a national asset, Scotland's footpath network has no overarching managing body. Instead, its well-being is subject to the whims and bank balances of a whole heap of disparate landowners, public bodies and charities. Many of these interests are now represented by NatureScot's Upland Footpath Advisory Group, who have been doing good work for a number of years to bring together some sort of coherent management strategy for the ongoing upkeep of Scotland's upland footpaths. In 2017, SNH (as it was then) published figures showing that users of upland paths contribute £110 million to the economy every year (by comparison, deer stalking, you'll remember, brings in around £50 million, and driven grouse shooting a further £25ish million, and these are the industry's own estimates). SNH's upland path audit from 2017, which incorporated information from ten different organisations, gives a good overview of the cost and scale of upland footpath maintenance, and deserves to be quoted extensively:

> Around 700 kilometres of paths have been repaired and restored over the last 30–40 years and the study has identified a further 410 kilometres that are in need of repair. An initial estimate of between £27 and £30 million is required for capital investment to secure and enhance the benefits of these paths to the nation, its people and the visitors who come to enjoy Scotland's mountains. The investment would, over a ten-year period provide stability to a fragile industry and secure upwards of 40 skilled jobs in rural areas (not accounting for any multiplier effects). This investment would help to sustain and nurture the estimated £1 billion of value contributed by upland paths to the economy over the same period ... This study has also estimated, for the first time, the likely resource requirements and employment opportunities to sustain the management of Scotland's upland paths. This will be an incremental cost during the major repair phase which would reach approximately £400,000 per annum to maintain 1,100 kilometres of upland path and provide long-term skilled employment opportunities for at least 20 people.[6]

£30 million, to secure £1 billion in value? If the figures are startling, then it is because the great benefits created by our footpaths can be realised so *cheaply*.

The problem with footpaths, of course, is that they are constantly trying to not be footpaths anymore. The actions of people and the weather, particularly rain, can cause footpaths to degrade very quickly. Maintaining upland footpaths is truly a Sisyphean task, and Paul Bolton is the man on the ground managing the hundred-odd miles of footpaths that crisscross Mar Lodge. He's in the right job: Paul is a prodigious walker and scrambler. Like the Victorians, he has a prowess for the plod. I'm a fairly quick walker, but Paul is one of the few people I would struggle to keep up with on the hills. He's been part of the Mountain Rescue Teams in both Braemar and

Torridon, the heroic band of volunteers who give up their time to help those in need on the hills. His time in the hills has given him an instinctive 'feel' for the ways that footpaths work, and the way the Trust needs to work on the footpaths.

'There's always more to do, it's one of those Forth Road Bridge kind of jobs,' he says. The Trust has just finished its upland footpath audit, which involved assiduously mapping and assessing every metre of its 400 kilometres of footpaths across Scotland. It was, he says laconically, 'quite a big piece of work.' Paul's prowess for the plod is not limited to walking – he's also extremely good at plodding through the meticulous planning and execution of Mar Lodge's upland footpath work.

Paul's next, quite big, piece of work is a five-year project which aims to tackle all of Mar Lodge's footpaths' problem areas. The idea is that a big, strategic push will keep costs down in the long run. But this project isn't cheap: the total cost will be around £1 million, and it will all be paid for by Trust funds. In the five-year period that the project is running, Mar Lodge will expect to be visited by around half a million people. That's £2 per visitor, spent on footpath upkeep. That £3 for parking at the Linn of Dee seems quite reasonable now. I must have looked surprised when Paul mentioned the word 'million'. He chuckles, 'The good news is we came in under budget in 2019.'

It's a huge task. Paul lists off the places where work will be taking place in the project: 'Lairig an Laoigh, Beinn Bhreac, Glen Dee, Coire Odhair, Macdui, the Braeriach plateau, the Devil's Point, Beinn Bhrotain, Beinn a' Bhùird, Carn na Drochaide, Bynack, Carn Liath, Carn Crom, Derry Cairngorm, Etchachan, Luibeg, Sron Riach. And of course, we did Carn a' Mhaim and Beinn Mheadhoin this year.' And this is all on top of the regular maintenance work. Every year footpath specialists walk the site, particularly the areas prone to damage, and keep up a continual effort of small, quick fixes where they are needed. The idea is that a small amount of work today, to clear a drain for example, can save thousands of pounds' worth of work further down the line.

Paul next ticks off the usual difficulties of footpath work. It's physically demanding – footpath workers need to be seriously tough. It's difficult to find enough people capable of doing this incredibly skilled, difficult work. The Trust has its own in-house specialist footpath team, but there is more work across the Trust's land than they can manage themselves. Much of the work is done by contractors, 'and there's only a dozen or so of them to choose from.' Remote terrain makes access difficult, and the Scottish weather makes life even more difficult. Then there is the increasing footfall on the paths themselves. With better kit, more knowledge, more roads and better access to the hills come more people. Paul is seeing a trend of increasing footfall across the hills. 'This is great, so long as it's managed. These are really fragile environments. People want to keep to the footpaths, because most people are aware of the damage they can do to fragile plants and soils, and let's be honest, walking on the paths is a lot easier than going off-piste. But as land managers, our side of the bargain is to keep them in good order.'

However, working at Mar Lodge brings extra difficulties. Material is usually sourced onsite, but even then, some of it has to be helicoptered around. 'And when you start getting helicopters involved, the costs really ramp up.' Then there are the eagles. 'We have to wait until the breeding season is finished before the choppers come in, or before we can work on paths that are close to their nests. That can really shorten the time we have to do the work.' Then there's the weather. Nothing can really be done between December and March. 'Obviously it's a pain for the teams to be working in poor conditions. A spell of bad weather can really slow things down.' Then there's avalanches. 'They can wipe out thirty grand's worth of work in a few seconds.' Even footpaths are beginning to be affected by climate change. 'We're beginning to see more extreme weather events. Flash flooding and heavy rain can wash away even new sections of paths.'

While many visitors to wild land areas will remain unaware of the path below their feet, at least until it gets into a sorry state of repair, there is a growing appreciation for the footpath networks that criss-cross the country, and indeed the skill and finances required to maintain them. The Trust's ongoing footpath fund is one of their most successful ever fundraising campaigns, which is just as well, given the challenges of managing 400 kilometres of footpaths in some of the remotest parts of Scotland. For people wanting to get their hands dirty, the Trust also offers Thistle Camps, week-long working holidays which mix practical conservation and days off to enjoy the hills. Paul and the team have started offering tailored 'Mountain Thistle Camps', which mix ranger-led walks up the mountains with footpath maintenance work.

The good news, then, is that the word seems to be getting around that, as far as environmental conservation is concerned, footpaths are a (comparatively) easy win in a sector used to tackling very complicated problems. Here are projects where the impact of one's cash can directly be seen under foot. These works are a rare example of an uncomplicated, uncontroversial piece of conservation being put together by a huge variety of organisations for the benefit of everyone.

15

Dotterel

> I only wish that I now understood as much about them as I then thought that I did.
>
> Desmond Nethersole-Thompson, *The Dotterel*, 1973

You'll remember Skitts from the chapter on sphagnum. That day, as we drove home, he told two anecdotes which are worth repeating.

The first concerns dotterels, the ephemeral, beautiful waders that haunt the high mountains for a few short weeks in summer. In 1988, Skitts and Adam Watson were on the Mounth, studying a male dotterel with a colour-ring. Dotterels are migrants – they come to the high and cold places of the world to breed, and then sensibly head off to Africa for the winter. The colour-rings help to monitor these mammoth movements – people who see them in the field can report their sightings to national monitoring programmes. Skitts and Watson sat down a respectful distance away from him and watched him for half an hour, waiting for the bird to lead them to a nest. Just as he was planning to do exactly that, he disappeared into low cloud, and that was the last that they saw of the bird. But it wasn't the last that anyone saw of it. A month or so later the bird turned up in Norway, chancing his arm at breeding in both Scotland and Norway in a single season. A few months after that he was shot dead in the Atlas mountains.

The second story harks back to the 1970s and concerns Desmond Nethersole-Thompson, who watched dotterels and really started off all the stuff that goes on in this chapter. A young Skitts and an old

Nethersole-Thompson were in the Glen Derry woods, working on Scottish crossbills. 'I'll just go check on an eagle nest,' says Skitts, leaving Nethersole-Thompson waiting, wrapped up in his big parka jacket and stalker's hat. Eagles and crossbills are both early breeders, so it came as no surprise when it started snowing. Back comes Skitts, three hours later, to find Nethersole-Thompson still watching the crossbill nest where he left him, having seen some interesting bird behaviour, and with several centimetres of snow tottering up on his hat.

The stories serve to illustrate, if not the profundities of the universe, then at the very least the hardiness of both dotterels and those who would study them.*

A further word on Desmond Nethersole-Thompson. Desmond, and indeed the entire Nethersole-Thompson clan, are real conservationists' conservationists. He possessed a knack for detailed observation, a mania for collecting data, and extreme hardiness of a type rarely attainable in our Gore-Texed days. He wrote the book on dotterels, literally, but only after studying them for thirty years, spending weeks at a time camped out on the Cairngorm plateau.[1] His legacy lives on, literally. Professor Des Thompson, his son, is currently head of natural research with NatureScot.

So Nethersole-Thompson is a conservationists' conservationist, and dotterels are proper birders' birds. They remain relatively unknown beyond those who make a point about knowing about birds. This is a shame, because they are quite lovely, and deserve a bigger audience. They are quirky, charismatic creatures, with cream eyebrows, a russet front, short stubby bills and loping long legs. Their cachet comes from three places: rarity, which is compounded by their proclivity for living in places which are a bit hard to get

---

* Pressing this point home, Skitts tells me, 'It was the eagle work that led both Adam [Watson] and myself to look at dotterel. What happened was we were both young fit men and used to take "short cuts" over mountains between eagle sites.'

to; their beauty, which is compounded by their proclivity for living in beautiful places; and their endearing tameness, which is compounded by their proclivity for living in wild places. When I moved to the Cairngorms the dotterel was the bird I most wanted to get acquainted with.

'There is only one first time,' wrote Nethersole-Thompson, for seeing your first dotterel. His first time was in 1933. The second time, he wrote, 'the Cairngorms soon reduced the bold and bumptious man down to size.'

Beyond the occasional outlier, in Britain dotterels only breed on the great whaleback mountains of Scotland. For a southern Englander, dotterels were as much the allure of the Cairngorms as the pinewood birds, the capercaillie, Scottish crossbills and crested tits. Growing up, watching birds in Devon and London, I never really thought that I would actually see a dotterel. They possessed the exoticism of the far north. They had the pang of adventure, hardship and conquest, difficulty and reward. They were to me an impossibly rare jewel, over the hills and far away.

The first time I saw them I was walking up Beinn a' Bhùird. It was late June. I set off early, walking the long track of Glen Quoich through the pinewoods. This was in 2016, when the young regenerating pines were putting their heads above the heather in truly big numbers, and the birches were a little further behind. Amazing, indeed, the amount of change in only the last few years. I climbed the old vehicle track, which is now a footpath, out of the pines. This was a remarkable piece of conservation work in its own right – the first removal of a vehicle track at high altitude in Britain. I started watching in earnest from the 800-metre contour, scanning the ground, waiting for a plover shape to break the skyline. And then, miraculously, one did. It was a male with a chick.

I kept my distance – dotterels are Schedule 1 birds, which means it is illegal to disturb them either on the nest or when they are with young chicks, unless you have a licence to do so. Famously, the females leave the males to do most of the work of brooding the eggs and rearing the chicks. I watched them grubbing about for a while, investigating their surroundings in fits and starts, occasionally reaching down for a morsel of cranefly. Then they disappeared over the horizon, and that was that. That is part of their allure – they come and go, and if they don't want to be seen, then you won't see them. But they can be very confiding creatures.

Now, as a resident in the Cairngorms, dotterels have a different allure, one not of exoticism but of familiarity. Where once they formed part of my mental landscape of a faraway place, now making the pilgrimage up to the high plateaux to spend some time with them is as much a part of my annual calendar as Christmas. And yet, like hearing the first willow warbler of the year singing outside my window, seeing dotterels is a poignant affair, one that comes from the knowledge that they come and go. It is a love borne out of transience, a nostalgic love made painful by the knowledge that they are only here for a few weeks. I think of them a lot in the long, dark winter. I am not alone in this: Nethersole-Thompson referred to dotterel song as 'acutely nostalgic': 'No one who has lain in a tent at night in the misty hills ever forgets the little drips of tinkling sound, always beckoning and often tantalising.'

Like a person, a place cannot be understood in isolation. It is only in understanding the creatures which call it home that you can come to something like an understanding of what a place means. Birds lend a place its character, while a place's character defines its species assemblage. Birds are the embodiment of the places they live – the world made flesh. It is all too easy to fall into the trap of thinking of ecosystems as complex but unthinking machines, working like clockwork, a mere backdrop to human affairs. But the comings and goings of our birds, should you care to notice them, provide a deep connection with one's environment, a reworking of the mechanics of one's perceptions of the environment. The act of noticing the creatures we share the land with is an expression of what it is to be a human part of a complex ecosystem.

So – rare, interesting, beautiful. But that nostalgia, as happens so often, is now taking on a more painful hue. Unsurprisingly, the dotterel is declining.

Nethersole-Thompson watched dotterels with his family most years for three decades. They would camp up at 1,000 metres altitude on the Cairngorm plateau and in the Grampians. They found out remarkable things: the Grampian birds, for example, would nest earlier than the Cairngorm birds, and enjoy better breeding success. This is because the Grampians are lower and more fertile than the Cairngorms, with a greater covering of mosses. They worked out when the birds arrived and left, their breeding density, which plants they associated with, how long they would incubate eggs for, how long it would take the chicks to reach maturity. They followed a couple of individual dotterels over several years, growing close, mutual bonds between the birds and their watchers.

Beyond their beauty, and mere scientific curiosity, dotterels are hardly ten a penny in Scotland, so they deserve to have an eye kept on them. For the last fifteen years, a small, ragtag bunch of professionals and amateurs has spent one day a year following in the footsteps of the Nethersole-Thompson clan, doing just that. July 2019, and this year there are ten of us, a mixture of estate staff, volunteers and local birders. We muster excitedly in the courtyard at first knockings, and Shaila gives a brief overview of what we're up to. We drive through the Quoich pinewoods, stopping briefly from time to time to look at crossbills and hen harriers. Newcomers admire the regeneration crowding the vehicle tracks. It's a good time to catch up with the gossip from the conservation world, from up and down Deeside and further afield. We stop at the final outpost of the track and start the two-hour trudge to the study site. It's a very pleasant trudge, accompanied by eagles and ravens, our feet skipping over dwarf cornel and trailing azalea, but a trudge nevertheless.

There's a butt beside the path, a vestige of a time when driven stag shooting was still a thing. The butt remains, and it is good shelter from inclement weather. But we're still a couple of kilometres from the start point, right up on the plateau proper. To ensure replicability, exactly the same area is surveyed every year, so GPS points and compasses are used.

Beinn a' Bhùird, the table mountain, is perfect habitat for dotterels. As the name suggests, it is a long, wide, high plateau, stretching five kilometres, joining up with the Yellow Moss to the west and Ben Avon to the east. It is a huge area of potential dotterel territory, and big enough to be undisturbed by people. The surveyors fan out between two flankers tasked with keeping the line true. The survey itself is a police line, which takes in around ten square kilometres of the highest upland of the massif. Alongside dotterel we record ptarmigan and snow buntings, and the occasional ring ouzel, wheatear and golden plover. There is also a speck of time for casual botany: parsley fern, three-leaved rush, moss campion and thrift all make appearances.

Radios and GPS loggers are liberally sprinkled between the surveyors to keep us on the correct path and make sure everyone knows what everyone else is up to. The line stops and starts as people log birds. Shouts and mimes go up and down the line, ensuring that birds are not double counted.

I'm a flanker. I stop at the summit cairn of the North Top and the police line swings around. This is a particularly difficult affair, as people need to walk at different speeds, with those in the centre staying still while a full 180-degree sweep is enacted. It's a warm, bright July day but I still need my ski gloves, making juggling radios, GPS, notebooks, binoculars and cameras difficult. There's a snow bunting singing next to me, and a female feeding its young a few metres away. It's enough time for me to become concerned by the lack of dotterels found thus far. Eventually, the line sets off again, and we take in another couple of kilometres.

Lunch is taken overlooking the corrie. This is the halfway point. Numbers are taken and written in duplicate. No dotterels have been seen. The corrie is magnificent, with a lochan and a smattering of snow, but it is cold, so no one is keen to linger for too long. One of the nice things about counting dotterels in this way is that it can be done in reasonably poor weather. The survey would only be called off if the weather was so bad that disturbing the birds would cause them the potential for harm. In 2018, a year when it barely rained for three months, and it was so hot that the only way to bear working some days was to occasionally throw oneself into the nearest burn, the party somehow contrived to find the only misty day in July to take on the expedition.

The survey continues. The ground is rougher this side, more boulders and scree, and steeper edges. This time I'm the outside flanker. I pick my way through a boulder field and snow patches. A hen ptarmigan calls softly to her hidden young, and sulks away from me. Meadow pipits somehow make their living up here at over 1,000 metres altitude. We struggle to keep the line. Somewhere above, people have found a dotterel, the radio crackles, with a chick. I'm relieved. As we're finishing up the line, another ptarmigan flushes, this time giving the broken-wing display. Six chicks erupt from almost underneath me. They're tiny balls of fluff, but somehow they're already capable of flying strongly.

A final line takes in slightly lower ground. This area tends to be used in the harder springs and less in the warmer years. Today, I see ptarmigan and a juvenile wheatear here, but no dotterel.

It is a nice day. It is, I think, what people think of when they think of bird surveys: slightly odd but mostly pleasant people going out of their way to look at obscure creatures. It has a ring of familiarity about it, like going to church. With the exception of the GPS coordinates, it is roughly how ornithology has been done since the beginning of ornithology.

But the 2019 results are troubling. Fewer dotterel are recorded than ever before, and there's a low count of ptarmigan too. The wet spring has taken its toll up here. Dotterels and ptarmigan are remarkably resilient and long-lived birds. Populations can survive a few years of poor weather. They are also capable of exploiting good-weather years. What they cannot survive is continuous poor weather over longer periods, and sustained competition from other species. The birds will have arrived, and settled on eggs, only to be forced to abandon them by days of heavy rain and snow.

Climate change brings not only more regular extreme weather, but also a loss of the regular rhythms that make ecosystems function. Dotterels are reliant on craneflies as their main food. They time their breeding to coincide with the main emergence of craneflies on the mountains. But climate change is pushing the emergence of craneflies earlier and earlier, and dotterels are struggling to keep up. Even in years without extreme weather, the odds of them breeding successfully are being tipped against them.

We leave the tops, retrudge our steps, and spirits are noticeably low.

Nethersole-Thompson saw disturbance from people and their hill tracks and mountain bikes as the dotterels' main concern. 'The Highland weather is soon likely to be the only factor favouring the dotterels,' he wrote. Now, even this seems uncertain.

In the grand scheme of things, dotterels are common enough birds. They live in the Arctic from Norway to Kamchatka in eastern Siberia. The Scottish population is a mere dot, interesting only for its quirk of being further south and further west than most of the main populations. For all its size and grandeur, the Cairngorms are again shown to be a speck in the grand scheme of things.

But this does not mean that the Scottish dotterels do not have value. For one thing, they are *our* dotterels – are we not entitled to our own wildlife? Does this population not have its own rights to exist on this land, as they have done for the last 10,000 years? For another, the Scottish hills will be among the first to be hit by climate change and anthropogenic destruction. The loss of our dotterels is the harbinger of greater losses elsewhere. Indeed, populations in Scandinavia are also already suffering.

Surveying all of the dotterels in Scotland is a big piece of work. Nevertheless, it has been done several times over. In 1987–88, the

Scottish population was estimated at around 1,000 males.[2] In 1999, when the next national survey was done, that number was reduced to 747 males.[3] In 2011, it was down to 423.[4] We can assume that it is now lower still. The Mar Lodge study site provides some local context to this national problem. Here, numbers are falling, but not as quickly as 2019's torrid count would suggest.[5] The Cairngorms may not have the most productive dotterel habitat – that accolade falls to the Grampians – but the region can claim to be the largest of the dotterels' Scottish haunts. If they cannot survive here, then they are in very serious trouble indeed.

A new, extremely thorough, paper has looked at what is causing us to lose Scotland's dotterels.[6] It found that dotterels are retreating 25 metres up the hills per decade. At this rate, it will be a mere matter of decades before dotterels simply 'run out of hill' in Scotland, and have nowhere further to retreat. The species appears to be retreating to its population strongholds, with geographically isolated dotterel pairs more likely to be lost than pairs nesting nearer to other pairs. The paper also highlighted the problem of increased nitrogen deposition in the Highlands. Nitrogen is essentially a fertiliser, and it tends to reduce the extent of arctic-alpine plants in favour of more vigorous species, reducing the suitability of the habitat for montane birds. And there are other problems in their wintering grounds, in which they spend the majority of the year. Breeding success is linked to rainfall and land-use patterns in the Atlas mountains.

What do we do to save the Scottish dotterels? The sad answer is that there is little we can do on the hills. Dotterels may be tame enough creatures, but they prefer to be left to themselves. Reducing the impact of walkers, dogs and mountain bikes is pretty much all there is to be done. To save Scotland's dotterels, we need instead to fight climate change. If you want to help save the dotterel, the three best things you can do are reduce your meat consumption, stop taking flights and plant trees. But ultimately, this is a societal problem. If you want to see dotterel in Scotland in fifty years' time, then you need to campaign for large-scale global political change.

In this intractable situation, naturalists find themselves in the awkward position of watching the slow decline of the natural world, unable to do much but continuing to shout that Something Must Be Done while not enough is. It may turn out to be the case that this study is nothing more than a document of the extinction of the dotterel in Scotland.

Dotterels are among the world's hardiest creatures, but they have little defence against climate change. Nethersole-Thompson finished his book on a note of, if not optimism, then at least affection. As is so often the case in the age of the Anthropocene, his words harbour new, tragic undertones. 'Whatever happens in the future,' he wrote, 'these lovely and confiding dotterels will always call and hold us. We remember them with affection.'

16

Downy Willow

It's late February, and it's a bizarrely warm 14°C. I systematically peel off all my winter layers as we work our way up one of the burns that falls off Carn Bhac. Shaila and I are at the southern outpost of the estate. Carn Bhac is the only Mar Lodge Munro south of the Dee. It's a handsome hill, big and green and mossy, with the classic whaleback profile of the Grampian hills and a predilection for avalanches on its northern scree face. Not this year though. In fact, it is the lack of snow that has drawn us up here, as it is allowing us to steal a march on the year's fieldwork.

We stop at the top of the burn, about 800 metres altitude, and look back across remarkably snowless Cairngorms. By now I'm in a T-shirt. We've passed a couple of paltry snow patches. They cling to the innermost ledges of burnsides, away from the direct attention of sun, wind and rain, a natural receptacle for drifting snow coming in from the south-west. They are soggy and crumpling around the edges, not long for this world.

We head back the way we came, stopping as we reach a small bundle of twigs. It is a scrappy thing, dangling over the edge of a small waterfall. It is not much to look on at the best of times, but on a grey February day, the bunch of leafless twigs is distinctly unimpressive. We examine its spent leaves through binoculars, determining that these are indeed the twigs we are looking for. I pluck up a bit of courage and dangle myself over the ledge to collect a few samples. I'm not the first to do this – a deer has been nibbling at its edges. We take a few leaves back, marked with the GPS coordinates, and head onwards. We've about a dozen scraps of twigs to look at in total.

The twigs are largely, but not exclusively, from downy willow, *Salix lapponum*. It's among the rarer plant species to occur on the estate. Downy willow forms small bushes, rarely more than waist high, with leaves very vaguely resembling sage. A quick look at the leaves, even when they are old and desiccated, will show you that it is well named – they are covered in a downy fuzz which keeps the cold of the Arctic out of them. It has been nearly twenty years since anyone has made a systematic search for them in the area, and it is high time that was remedied.

It is exactly the sort of task that certain types of people find hugely enjoyable: a big, landscape-scale treasure hunt for rare and exciting things, scrambling about in a rather lovely series of waterfalls into which no one ever really goes. We have the handwritten notes and scrawled maps from the previous, pre-digital search. We spot woodcock and grouse, and lots of stags. It is an enjoyable way to spend a day. All in all, we fare as well as our predecessors. We put GPS coordinates to their handwritten notes, in preparation for further work later in the year. For our efforts we are rewarded with views of a literal handful of sticks that one of us could have happily carried off the mountain – barely enough to get a fire going.

The reason for this paucity is obvious. Deer dung carpets the glen. Eroded tracks, covered in deer footprints, lead down to each willow. As soon as plants put on growth that is even moderately in reach of a hungry mouth it is being nibbled. This is why the only place we find the willows is overhanging burns, cliff edges and waterfalls. Downy willow is joined in some of the larger gullies by other trees, rowan, downy birch, grey and eared willow, all refugees from the deer above. One or two slightly more accessible downy willow bushes are instead protected by snow patches – patches which are getting patchier. Most patches are too isolated from other willows to successfully reproduce. It is, as Shaila glumly notes, 'desperate'.

For all that, the persistence of these bushes is heartening, less for their ecological function and more for what they represent. They are the key to a lost ecosystem which once stretched across the Scottish uplands. They are some of our montane willows, one of our great unknown national treasures, and they are nearly all gone, lost to centuries of human indifference and the nibbling of too many grazing animals. The remaining montane willows, shut like Rapunzel in lofty towers of stone, are distillates of an alternate reality in the Highlands, one that was lost, but one that may return.

Given the extreme rareness of the montane willows in Scotland, it takes some imagination to believe that only a few centuries ago a significant proportion of the mountains of Britain were carpeted with a habitat that has simply ceased to be. So let's take a walk up an imaginary, ecologically intact mountain. We'll start at sea level, and we're quickly into a mixture of woodland, thorny scrub and more open areas where the grazing is good. Here, the trees are oaks, ash and hazels, with big willows in the wetter areas. As we clamber up the mountain, through this dense stuff, the oaks begin to thin out, to be replaced by birches, rowans and pines, hardier trees which demand less fertile soil. We continue to climb, and even these start to thin out, as the conditions become colder, windier and maybe wetter. Soon the pines are getting smaller, crumpled over by wind, snow and exposure. This is the timberline, the altitude at which trees no longer grow straight and tall. The twisted, stunted pines are known as Krummholz pines – the German for 'twisted wood'. The birches, too, are beginning to become stunted by the altitude, but by now, in the places with slightly richer or wetter soils, scrubby willow species, eared, grey and creeping, are growing as dense bushes. Juniper, which is an understory plant further down the mountain, is beginning to take a more major role in the landscape. As we continue getting higher, new species begin to filter in, like dwarf birch and downy willow. By now even the hardy birches are beginning to struggle to grow. Keep climbing, and this scrubby mixture of birch, Krummholz pines, juniper and several different willow species gets lower and lower to the ground. Finally, we top out, and are walking through low shrubs like heather and northern blaeberry. This is the final edge of the 'scrubline'. Above this is a mixture of rock, grassland, bog and montane heath. Above this, at the highest tops, lichens and mosses will be the dominant species.

The places where different ecosystems blend with one another are called ecotones. The demands of different trees in different places vary, leading to a mishmash of ecotones as we ascend the mountain. All of this diversity creates a huge number of niches for different species to fill. There are loads of different factors that go into determining where a 'natural treeline' will occur on a mountain. Two of the most important are latitude and altitude. Put simply, a higher mountain at a lower latitude will have a similar species composition to a lower mountain at a higher latitude. This is why a relatively low-altitude mountain like Ben Lawers has a similar range of plants to the far higher mountains of the Alps. This is also one of the reasons why Scotland has particularly interesting botanical things going on, as it harbours a range of plants found both in the Arctic and the Alps. Temperature, precipitation and wind also play a big role in 'stunting' trees at high altitude. In the east of Scotland, where the climate is more continental and less oceanic, the treeline tends to be higher, even though it is often colder and snowier here. In fact, there are several hundred metres of difference between the treelines between the east and west of Scotland, and indeed the north and south. Then there are local variations, depending on the underlying rock, the drainage, soil conditions and shelter. In the Cairngorms, the 'natural timberline' of Scots pine tends to sit around 600 metres altitude, but the highest recorded pine at Mar Lodge, a tiny, twisted, stunted knee-high specimen, is around 900 metres altitude.

Scotland is one of the very few mountainous regions in the world which does not possess an extensive montane scrub zone. Here, this entire ecosystem and the creatures that it supported have been systematically erased by centuries of burning, clearing and grazing by deer and non-native sheep and goats.[1] The Scottish

Semi-natural Woodland Inventory, compiled in 1988–89, found just 27 hectares of semi-natural woodland surviving above 700 metres altitude, with a further 341 hectares between 600 metres and 700 metres. It is widely held that just 600 metres of 'natural treeline' remain in Scotland, at Creag Fhiachlach on the Speyside side of the Cairngorms. A study from 1997 concluded that Scotland could support 700 square kilometres of montane willow scrub – just one type of many montane scrub ecotones.[2] As it stands, only around one square kilometre remains, and this is spread out across the country, a diaspora of tiny clumps of habitat. But what is certain is that there clumps no longer form part of a thriving ecosystem. They are instead a naturalist's curiosity: a museum piece. As a habitat, montane scrub is pretty much functionally extinct in Scotland.

It is hard to say when it really, truly disappeared. But in all probability montane scrub persisted as a fairly major habitat at Mar Lodge until close to the Clearances. The word for 'willow' in Gaelic, *seileach*, will be well known to anyone who has pored over maps of our mountain areas, as will *chaorainn* (rowan). On Mar Lodge there is a Cairn nan Seileach, the willow hill. Ten miles away, there is a Coire an t-Seilich, willow corrie, and below that the Allt Coire an t-Seilich, willow corrie burn. Both of these sites sit above 600 metres altitude, well within the range of montane willow species. Gaelic did not reach Deeside until the (very) early middle ages and remained the language of the Mar Lodge area until the late nineteenth century. In the years since the loss of the montane scrub zone, its situation has only become worse. There is now more grazing on the mountain tops, not less, and climate change is pushing the woodland zone further up the mountain.

It is also hard to say how long these shattered sticks have left to live. Their situation has changed very little in the last twenty years, when the previous survey was done. Given its tortured history, what is really remarkable is actually how prevalent downy willow remains. Anyone who has worked with trees will know that willows are pretty bombproof: fast-growing, quick to set seed and colonise bare ground, happy enough in wet soil. Despite the best efforts of centuries of nibbling herbivores, montane willows persist in a couple dozen locations across the Cairngorms, with maybe a third of these occurring on Mar Lodge land. All the sites possess the same qualities: high, snowy, out of the reach of deer. Say it quietly, but they might just have persisted long enough for us to get round to saving them.

February's preliminary visits may not have been particularly encouraging, but as it turns out, 2019 may just be a watershed year for the Cairngorms' montane scrub. There are two reasons for this. The first reason is injections of cash to finance montane scrub restoration work at both Mar Lodge and the giant Cairngorms Connect partnership next door. In 2018, the Cairngorms Connect project received £2 million in funding from the Endangered Landscapes Programme. A significant chunk of this money will be used to fund montane willow conservation. Meanwhile, Mar Lodge received £100,000 to assist with woodland restoration work from the People's Postcode Lottery.* This cash injection allows work to go beyond the theoretical and into the practical. There's strategy afoot.

The other reason is Adele Beck, a specialist ecologist in montane willow species. Adele and her team have spent four field seasons mapping the extent of montane scrub in the Cairngorms as a baseline survey for potential montane scrub restoration projects. She's worked to the same methodology across Cairngorms Connect and Mar Lodge land, and has come up with the some of the largest, most detailed vegetation surveys ever undertaken in the Cairngorms.[3]

The results are fascinating. Her team's work shows what is present, and what isn't, and how much of it is being eaten, and where. It shows that montane scrub species are generally more common than conservationists thought they were. Two species in particular, dwarf birch and juniper, are far more common than was previously

---

* Pause to note that the survival of some of our rarest natural assets will end up being paid for by the whims of charity.

thought. The survey also finds reasonable amounts of a couple other important but less specialised montane scrub species: tea-leaved willow and eared willow. It shows that montane scrub has been benefitting from the reduced deer population in the regeneration zone, though probably not to the same extent as the pines down in the woods. This is vital evidence – it shows that further conservation work beyond maintaining low deer numbers will be unnecessary for these species to return in a big way to the mountains.

In 2019 it is time to move on to the next phase and look more closely at the two species which Adele's survey has flagged up as being the most in need of a helping hand – whortle-leaved willow and downy willow. The team follows up all historic records held for these species, taking seed and cuttings, counting the total number of plants at each site, checking them for browsing. It is tough work, often involving specialist climbing equipment to get to the least accessible ledges, and it is not helped by the wet weather of the 2019 field season. To make matters more difficult, each plant will only flower and set seed for a week or so each year, so timing is vital. Get it wrong, and you might have wasted an entire day simply walking to and from the specimen.

More generally, montane scrub seems to be popping up everywhere in the regeneration zone. Alongside the rarer stuff, we're noticing more and more eared and grey willow in the uplands. These commoner species often get short shrift in conservation circles – they are, perhaps, the forgotten montane scrub species, and will likely be doing a lot of the heavy lifting in newly emerging scrubby areas. I find five 'new' patches of tea-leaved willow myself, while out and about looking for other things. In early May, I walk high up in Glen Derry with Gavin Legge from Forest Research. He's here on holiday, helping out with the raptor monitoring work. We walk for a couple of miles through scree and shallow soils on a steep gradient, about 750 metres up, with emerging montane scrub almost constantly at our feet. We're both astounded by the amount of juniper. The bushes have been hanging on, prostrate, for centuries, perhaps. Now, they are exploding outwards and upwards. Rowans and eared willow grow out from among the juniper – its spiky leaves are providing protection for these palatable species from deer and mountain hares. One rowan, somehow, is making its living out of the top of a boulder. And pines, Krummholz, contorted twisted formations, join the fray. It is the most extensive piece of montane scrub I have seen on the estate, and among the most exciting bits, for my money, in the country. It is indeed the most exciting walk I have been on for a while. Gavin is as surprised as I am to see the emergence of a whole new habitat, lost for centuries. An unplanned, unintended consequence of reducing deer levels to protect the Caledonian pinewoods, several square kilometres of one of Scotland's rarest habitats are returning to the glen before our eyes. 'I've learned more in the last couple of years from projects like this than in my previous thirty years of work,' he says. 'In ten years, this glen will be unrecognisable.'

June, another bird survey, another early morning, another existential crisis. I plot a grid reference and satellites guide me off the path towards a small burn. Here is a patch of downy willow, unrecorded until Adele's team found it two years ago. It comprises two small bushes, no more than knee height, overhanging a peat-stained burn. The pale-green fuzzy leaves shine brightly against the browns and yellows and dull greens of the surrounding heather and grass. 'Well, hello,' I say to the bushes, as I have developed the unfortunate habit of talking to plants in the absence of people. 'Have you anything for me today?' Yes, as it happens; the bushes are producing catkins. They are curious things, long pale spikes, with the smallest hints of fluff developing about them. I pick a freezer bag full, maybe a tenth of the catkin crop, and am on my way, back to the bird survey.

By the time I reach the Hutchison Memorial Hut for a spot of breakfast I have forgotten about the willows. The snow is still lying in the upper reaches of the corrie, bridging the burn in places. Continuing up to Loch Etchachan I hear a bird calling – a grey wagtail. It's an unfortunately named bird. Yes, it is grey, and yes it wags its tail, but it is also one of the most beautiful creatures in Scotland, a flash of yellows, whites, greys and blacks, a lover of watercourses and a highly characterful thing. A young chick is begging for food while the adult contorts itself plucking insects from the air, its overlong tail a foil and balance.

There, to my astonishment, below the wagtail antics, is another patch of willow. It's huge, as far as montane willows in Scotland go, glossy green, covering maybe three square metres. 'Which one are you then?' I ask. Identifying mountain willows can be tricky, particularly for a jack-of-all-trades ecologist like myself. I take photos of the leaves and twigs and catkins. The snow-slicked gravels it is growing on give way under foot – this area, high, sheltered and north-facing, holds its snow late. As I slip, I notice another two plants, seedlings from the larger bush. The snow protects the willow from grazing deer, but also hides it from the view of people for more than half the year. But this can't be the only reason that no one has spotted it before. Like the downy willow further down the valley, this willow is growing, actually *increasing* in size. I have a final look. Had it not been for the wagtail I would have missed it. Beyond the willows, Coire Etchachan is transforming. Krummholz pines are growing above the bothy; I record one at 800 metres. They are joined by rowan, eared willow and the occasional birch. Juniper is spreading out from the burnsides, growing in both height and number, reclaiming what some translate from the Gaelic as the 'juniper coire'. Other plants like globeflower and northern blaeberry join stands of lemon-scented fern, making the beginnings of a montane tall herb community, another extremely rare habitat in Scotland. The plants are pollinated by blaeberry bumblebees and peacock butterflies.

Back at the ranch, I check the databases and confirm a sneaking suspicions. This is mountain willow, *Salix arbuscula*. This species is rarer even than downy willow in the Cairngorms. This particular plant was recorded at this exact spot sixty years previously and had not been noted again since. The ecological implications are that there is more willow here than we thought, and it is increasing – in an age of mass extinction, a once-lost ecosystem is, perhaps, returning.

In July, Adele makes a discovery. A previous survey has misidentified one of our clumps of willow. The little bush is in fact mountain willow. By now, her work is almost finished. She has taken cuttings and seed from all of the estate's known montane willow patches. They are put into the care of Trees for Life's tree nursery at Dundreggan, where a specialist team begins the task of growing on the cuttings. In September, the Dundreggan team tells us which seeds have been viable, and which haven't. We now know which stands are capable of producing seed and which are not. It's a complicated business though – some of the downy willow seedlings are 'atypical', suggesting that they may be hybrid plants. This is quite likely; willows are notoriously liable to hybridise, and are more likely to do so in places where there are few individuals spread out over a large area. Adele has finished her report, and it's more good news – the whortle-leaved willow populations seem to be capable of looking after themselves. Meanwhile, Dr Aline Finger at Royal Botanic Garden Edinburgh, who you'll remember is also working with alpine sow-thistle, is working on sampling the genetic make-up of material from all of the Cairngorms' downy willow patches. Her work shows us which plants are related to which others, giving us a good idea of the genetic diversity of the Cairngorms meta-population, that is, all of the combined disparate populations within the Cairngorms. Her analysis shows that there has been gene flow between willows across the breadth of the massif. This result suggests that downy willow used to be far more common than it currently is.

The work continues. In September, Shaila and I take a couple of trips into the montane scrub areas, have a good scour, and find a further two 'new' downy willow plants. Further excursions reveal more patches of recovering scrub. In October we head up Glen Derry once more, this time with the assistance of Cameron, the beat keeper, Derek from the estate team and a dozen volunteers, again borrowed from John Muir Trust. As some of the first snows of the year vaguely think about falling from the leaden sky above us, we set up a series of two-metre-square exclosures around selected juniper, eared willow and downy willow plants. Each plant in an exclosure is paired with an unfenced 'control' plant nearby. This will give us an idea of how much herbivory is still taking place on the emerging montane scrub. It's hard work – we're up at 600 metres altitude and several miles from the nearest vehicle track. Small flecks of snow whirl around as we work. The ground is unforgiving. We return at dusk, cold and exhausted. It's a good day. I'm reminded of Keats:

> Therefore, on every morrow, are we wreathing
> A flowery band to bind us to the earth,
> Spite of despondence, of the inhuman dearth
> Of noble natures, of the gloomy days,
> Of all the unhealthy and o'er-darkened ways
> Made for our searching.[4]

Talking to a reporter (for montane scrub is becoming newsworthy), Shaila says, 'We were fairly sure that montane scrub was a missing habitat in this part of the Cairngorms. Twenty years ago you could have walked around and thought there was nothing here. Of course, it was all just lying low, dormant, waiting for the browsing pressure to ease. It was a complete eye-opener. Thank heavens we didn't rush in and plant.'[5]

When dealing with sensitive landscapes and rare creatures, the precautionary principle is an essential brake on well-intentioned but potentially damaging conservation actions. Good conservationists understand the need for any intervention work to be backed up as deeply as possible by good science and sound ethical judgement. The Cairngorms landscapes are hugely important for the country. There was a vital need to take a measured, cautious, scientific approach to any potential landscape-scale restoration of the montane scrub zone, and to ensure that any intervention was as low-impact as possible. Ultimately, this has paid dividends. Diligent monitoring showed that of all the species recorded, only one, downy willow, requires human intervention to ensure its long-term survival. More resources can now be targeted directly where they will have the most impact.

But the precautionary principle can also be used as a scapegoat for inaction. All too often, conservationists seem happy to sit back and watch on as creatures move steadily towards extinction, frozen from taking decisive, potentially risky conservation measures by the fear of failure, or worse, damaging the prospects of the species they are trying to save. Montane scrub is now rebounding across the Cairngorms, but downy willow will struggle to retake its rightful place in the landscape without intensive human intervention. Those tiny scraps of twigs and leaves are now among the best-studied individual willow plants in the world. It takes a leap of faith to get stuck in to landscape-scale willow conservation, but it also takes a reframing of perspective to accept that these last fragments of our montane willows are not just museum pieces to be admired but not touched, but living parts of a living landscape, which hold the key to a better, more biodiverse future for our mountains. The teams working with willows at Mar Lodge, Abernethy and Glenfeshie could have succumbed to the lure of keeping these places 'pristine', free from intervention, in their 'wild' state. This would have absolved them of responsibility, but also resigned the willows to their fate.

Montane willow conservation has now reached the magic point where key actors possess the knowledge, the ability, the funding and the willingness to take action to restore the species to its former glory. This has been a long time coming. Behind the scenes, a tiny, dedicated group of montane-scrub enthusiasts has been fighting the corner of this habitat for the last couple of decades. The willows at other sites, notably Ben Lawers and Corrie Fee, are twenty years ahead of the Cairngorms, with both sites hosting increasing areas of planted and naturally regenerating montane willow species, which are coming down from their inaccessible refuges and recolonising the mountainsides. For now, though, the willows in those places remain behind deer fences, their inaccessible burnside and cliff-edge refuges replaced by posts and wire.

The work to return montane scrub to its Cairngorms home is a hybrid affair, combining practices both ancient and hyper-modern.

The flowery bands that bind us to the earth are wreathed not just from cuttings and seed, but also from databases, spreadsheets, grid references, emails and genetic fingerprinting. The next step involves strategy, statistics, experimentation and green fingers. The team at Dundreggan is growing on hundreds of seedlings from Cairngorms genetic stock, which can then be planted out in the wild. These will be used to bolster the genetic viability of existing populations, and also create new, bridging stands of downy willow, which will link up the existing populations. They will also keep back a 'seed bank' and cuttings as an insurance policy, in case the worst should happen to the wild populations.

But simply planting out seedlings into the wild is unlikely to work. The team's research has shown that 'mob planting', using a couple thousand seedlings in a fairly small area, and surrounding them with sacrificial plants of upland birch (grown on from Mar Lodge seed), will give them the best chance of survival. As always, Mar Lodge management errs against deer fencing. But the willows in the southern part of the estate, where deer density is still high, will be protected by fences in the short term, their populations reinforced with seedlings from nearby.

The botanist sees the landscape in a different way to other people. In plants, the botanist sees great shifts of geology and grand tides of ecological change. The botanist sees not only what is present, but also what is missing. The botanist sees environmental destruction and degradation and climate change, but also regeneration, healing and, perhaps, redemption. Science, both in the lab and in the field, is teaching us new ways of working with the environment, of understanding our place within it and forging the beginnings of a new partnership between humans and nature. But of course, this relationship is as old as the hills: as Nan Shepherd wrote, back in 1943, 'To know, that is, with the knowledge that is a process of living.'[6]

Unobtrusive, missed by most, the remaining scraps of montane willows that drape themselves over our burns beg questions: What do we see when we look into the mountains? What should we see? What could we see?

Epilogue: You Could Have It So Much Better

Nature is perhaps the most complicated word in the language.

Raymond Williams, *Keywords:
A Vocabulary of Culture and Society*, 1976

What does an ecologist do all day? Mostly, it is chart the decline of our wildlife.

Amidst the gloom of yet another unseasonably warm January, when bushfires raged across Australia and hurricane-force storms from the Azores caused flooding across the UK, I came across a chink of hope in a recently published scientific paper. It was titled, not very encouragingly, 'The dynamics underlying avian extinction trajectories forecast a wave of extinctions.'[1] The paper synthesised underlying extinction rates in bird species across the world and concluded that extinction risk is increasing in bird species. The researchers found that during the past 500 years, about 187 of the world's 11,147 bird species are estimated to have become extinct. But the new projection suggests that during the next 500 years, three times as many – 471 species – may become extinct. So far, so depressing. But this line stood out: 'We very conservatively estimate that global conservation efforts have reduced the effective extinction rate by 40 per cent.'

Put simply: environmental conservation works, really well, but there is not enough of it to keep up with increasing levels of environmental destruction. For all the doom and gloom, we know how to save species from extinction; we just have to find the resources, collectively, to do it.

Conservationists have been fighting rearguard actions for as long as people have been calling themselves conservationists. We are, for the most part, a pessimistic bunch. But it is with increasing regularity that I am roused from my gloomy days by the unsettling, surprising tug of hope. A new wave of conservationists is going on the offensive for nature, aiming to restore huge swathes of our land to its former glory. They are forging new practices in environmentalism and Highland sport, taking what works from both and discarding what doesn't. It takes courage to face up to the failures of the past. Now, people are doing it.

Hope returns, as I travel across the Highlands. From Assynt[2] to the Borders,[3] and from Aberdeen[4] to the Western Isles,[5] native woodlands are returning to Scotland. In Speyside, our greatest woodland area is growing, from Abernethy in the east to Glenfeshie in the west, with joined-up management courtesy of the Cairngorms Connect partnership.[6] The western pinewoods of Glen Nevis, Knoydart, Loch Arkaig, Ben Shieldaig and Beinn Eighe are in the hands of conservation bodies working to restore them. In Glen Affric a hugely ambitious scheme, already thirty years and more in the making, is aiming to reforest a thousand square miles of land, creating a huge wildlife corridor linking Dundreggan, Glen Cannich and Glen Strathfarrar with the west coast, returning swathes of Caledonian pinewood and montane scrub in the process.[7] In Deeside the forests are expanding, from the Forest of Birse to Glen Tanar to Ballochbuie. Around Braemar, the Morrone Birkwoods are also expanding, while Invercauld, as traditional a sporting estate as you could hope to find, is reforesting hundreds of hectares of land. In the Borders, the Carrifran Glen is returning to life, courtesy of an ambitious, volunteer-led experiment in large-scale native woodland restoration, galvanising efforts to restore thousands more hectares of lost woodland in the Southern Uplands.

It's not just in the woods that life is returning. Bogs are being restored, non-native trees and invasive species are being eradicated. Meanwhile beavers have returned to Scottish lochs and burns after a 400-year absence. Cranes are breeding again in Aberdeenshire, entirely of their own volition. White-tailed eagles, red kites and goshawks have returned, helped on their way by supremely dedicated individuals. Red squirrels are expanding their range north and west. Pine martens are spreading south. Corncrake numbers have stabilised after over a century of decline. Ravens are returning to long-lost nest sites.

More importantly, people seem to be rewilding themselves. For every big, headline-grabbing project, there are hundreds of small-scale plans to return wildlife to our lives. Restored landscapes provide ecosystem services, saving the country billions of pounds in flood prevention and carbon sequestration, providing a valuable timber resource, while also generating income through tourism and sustainable sport. With every environmental project comes more information on what works well and what doesn't. Conservationists are refining and expanding their work accordingly, benefitting more and more people in the process.

For twenty-five years, the Trust has been fighting exhausting rearguard actions at Mar Lodge, as have conservationists across Scotland. Now, we are all seeing the results of their hard labour. These results are changing the ways that we think about our landscape, challenging long-held beliefs and teaching us long-lost truths about the nature of Scotland. Following these successes, and terrified by the peril our natural world is facing, a new generation of environmental conservationists are making their voices heard. Backed up by new science and standing on the shoulders of their predecessors, this new generation is louder, angrier and more ambitious than any that have come before.

Against a background of environmental destruction, climate change and increasing inequality, there are small rays of hope for the future of our wildlife. At Mar Lodge, we are only at the beginning of the story. But we could have it so much better.

A couple of years ago I was doing some fieldwork. I sat close in long heather. Being low and quiet and not moving, I didn't disturb a roebuck as it broke cover from the woods and wandered within about thirty yards of my little hidey-hole.

I knew perfectly well it was a roe deer, a young buck at that, with a small set of antlers. And yet something of its size and gait, as it struggled through the rank heather with long, loping back legs, dappled-dawn-drawn as Manley-Hopkins would have it, sent the faintest trace of a shiver down my spine. While the logical, rational part of me knew that I was looking at a roe deer, a completely different part of my brain over which I have very little control was certain that I was looking at a lynx.

It was very peculiar. In my work I'm fortunate enough to work with charismatic creatures on a regular basis. I've done all the things that nature writers get excited about. But this roe deer encounter was different. It was an extremely uncomfortable experience, a good example of cognitive dissonance, and a reminder that when it comes to the most important things in life, people are never entirely rational in their outlook. When faced with even the shadow of an apex predator, even one as small as a lynx-which-I-was-fully-aware-was-a-roe-deer, interesting things stir in interesting places in the imagination.

For better or worse, the concept of 'rewilding' is extremely polarising. For this reason, I should state, right now, that there are no plans to reintroduce lynx to Mar Lodge. Once more, there are no plans to reintroduce lynx to Mar Lodge. And yet, when I'm out and about working on the estate, the fourth most common thing I get asked is, 'What are you doing about lynx and wolves?'

But what exactly *is* a lynx? When you are brought up on a cultural diet of tabby cats and lions and not much in between they can be difficult to picture. Plus, they do not have the same cultural cachet as the wolf or the bear. They don't eat Little Red Riding Hood or serve up porridge. Lynx are solitary, highly territorial animals about the size of a Labrador, but also a bit lighter than a Labrador. They have short, stubby tails. They are lion-coloured, but there is some variation in their coats; some are patterned, some are not. Lynx are the sort of cat that is perfectly adapted for stalking through the Caledonian pinewoods in search of roe deer.

Not entirely sure that I could picture them properly, I went to see the Cairngorms' only resident lynx in their enclosure at the Highland Wildlife Park near Kingussie. They are beautiful creatures, with long limbs and penetrating predator eyes. They have charisma. They have the supple lines and taut muscles of pure-bred killing machines. A female came over to have a look at me, padding around the edge of the paddock. She sat, sphinx-like, looked me dead in the eyes. I looked at her. I saw claws and talons and intelligence. I saw that I could never truly fathom the depths of that creature. But when you look at them in a zoo what you really see is the fence. To imagine them in the wild is to imagine a completely different kind of woodland to that which currently exists in Scotland: one with claws.

Like most neologisms, 'rewilding' started life in an academic text, with a fairly narrow definition. Since then, use of the word has exploded, from being an extremely niche word in 1995 to just a niche word in 2005. Following the publication of George Monbiot's *Feral* in 2014, the high-profile reintroduction of beavers to Scotland and the vocal advocacy of organisations like Rewilding Britain, Rewilding Europe and Scotland: The Big Picture, the word is now bordering on the mainstream. Unfortunately, like most neologisms, it defies easy definition. It is used as both compliment and insult, as an innovative way out of our extinction crisis and a straw man, a byword for both progressive, ambitious thinking and idealism over practical possibility. Given the political power of the word, it is worth our setting out some ground rules. Let's take Rewilding Britain's definition:

> Rewilding is the large-scale restoration of ecosystems where nature can take care of itself. It seeks to reinstate natural processes and, where appropriate, missing species

– allowing them to shape the landscape and the habitats within. Rewilding encourages a balance between people and the rest of nature where each can thrive. It provides opportunities for communities to diversify and create nature-based economies; for living systems to provide the ecological functions on which we all depend; and for people to re-connect with wild nature.[8]

In Scotland, this means regenerating woodland, allowing watercourses to meander and work where they will, de-intensifying upland management practices like non-native forestry, drainage and muirburn, and reintroducing certain creatures made extinct by our own hand. It is a means of management that aspires to non-management but accepts that this will usually be impossible. It is an acceptance that nature can look after itself perfectly well, and that, as much as possible, humans should tread lightly in the landscape. It sees economic opportunities – new jobs in practical work and tourism, cost savings from reduced flooding and carbon capture, diversified opportunities for field sports. It is as much about increasing the quality of life of people as it is about nature.

The astute reader will realise that this concept of rewilding is not far off what has been underway at Mar Lodge for the last twenty-five years. The Trust has been regenerating woodland, allowing natural hydrological processes to shape the land, reprofiling plantations and de-intensifying moorland management. The Trust is even reintroducing lost species, though these are so far limited to plants like alpine sow-thistle. And yet the Trust is wary of calling its work 'rewilding'. It is still, I think, too political, despite its growing appeal across society, and Mar Lodge has been deeply stung by past criticism of its land management. Yet it is in places like Mar Lodge where rewilding, whether named that or not, is having the greatest impacts, all positive, on society.

Proponents of rewilding are quick to note that reintroducing large apex predators to kick-start natural processes is just one possible part of a whole range of actions and attitudes that constitute the 'rewilding movement'. Indeed, they further note, the idea of reintroducing wolves and bears has been pounced on by those who are set against landscape-scale change in Scotland, and its champions set up as bogeymen with whom to discredit the whole rewilding movement.

Scotland is actually quite a long way along the road of species reintroductions already. Sportsmen were the first 'rewilders', reintroducing red and roe deer, and capercaillie, to areas where they were lost. Mar Lodge deer were used to restock the Rothiemurchus deer forest for sport in the 1840s. Now the list of reintroduced species in Scotland includes the white-tailed eagle, red kite and beaver. 'Unofficial' reintroductions have included wild boar, goshawks and polecats. Other recent 'translocation' projects have involved red squirrels, black grouse, pine martens, polecats and golden eagles. In 2018, the beaver was 'officially' welcomed back to the Scotland, getting legal protection as a native species, after a 400-year absence. The wildlife of Scotland is not nearly as wild as people think. But that's by no means a bad thing – it shows that we need wildlife, and wildlife needs us, and that the distinction between the wilderness and civilisation is much blurrier than we are usually able to imagine.

The UK is one of the biggest countries in the world to possess none of its original large carnivore species. We have killed them all through hunting and habitat destruction. This has had a profound effect on our ability to even consider the possibility that the land wasn't specifically designed for our pleasure and dominion. It has fundamentally shifted our perception of the environment, pushing us to both idealise and fear large wild animals. It has had an equally profound effect on our ecosystems. As we have seen, the gap at the top has allowed smaller predators, foxes, stoats, crows, to flourish, which has had knock-on effects on interactions across whole ecosystems. It has allowed herbivores to flourish unhindered. By bringing back the top predators like lynx, we can help to restore natural processes, such as predation, anti-predator behaviour and the provision of carcasses that shaped our ecosystems for thousands of years – and can enrich them once again.

Given that I have spent a significant proportion of this book explaining the difficulties, trials and tribulations of the relationships between woodland and deer interactions, and explaining that the problem is one of a lack of natural predators, it is very easy to come to the conclusion that yes, chucking a few wolves and lynx in to some lonely glen and leaving them to it would save everyone a whole lot of bother. Indeed, for Mar Lodge, having a charismatic, flagship species such as lynx would provide an incredible boon, a draw for visitors in search of wild experiences and, perhaps, a major new source of income after a decade of funding cuts. Having also spent a significant proportion of this book explaining that our environment exists not just in the field but also in the mind, and the received wisdom of our ancestors, and the cultures and ways of life that humans tend to make for themselves, it is also very easy to come to the conclusion that such a project simply isn't possible.

If you want to talk to anyone about lynx in Britain, or indeed anyone in Britain about lynx, then you talk to Dr David Hetherington, woodland advisor for the Cairngorms National Park Authority. He literally wrote the book on the subject, *The Lynx and Us*.[9] David is very keen to manage expectations. Lynx are not necessarily the ecological silver bullet that many conservationists expect. But nor are they the sheep-savaging monsters that you read of in the media. David's simple argument is that the rest of Western Europe saw the extinction of lynx, just as happened in the UK, and that they are now living perfectly well with their newly returned neighbour. If other countries can do it, why can't we?

Conservationists are particularly drawn to the idea of reintroducing lynx, thinking that this will be more acceptable to people than the larger, pack-hunting, Little-Red-Riding-Hood-eating wolf. There are also good social reasons for reintroducing lynx. As a potential future predator of our exploding roe deer population, lynx would be an ally to foresters and assist Scotland in reaching its carbon targets by helping more trees to grow. Then there is the potential income from wildlife tourism. Helped along by reintroductions, lynx are recovering across Europe, responding to declines in deforestation, over-hunting and persecution as vermin. In fact, large, charismatic animals are returning across Europe, from wolves to lynx to beavers to bison to bears. The only thing stopping their returning to our shores of their own volition is, well, our shores.

It is uncertain when the lynx disappeared from Britain. They were certainly here in the sixth century, to which the most recent piece of lynx bone recovered in the UK has been dated. But it is hugely unlikely that piece of bone belonged to the last lynx to stalk Britain. The balance of probability suggests that they died out at some point much later. There is a chance that lynx may have lived on in the Highlands until as recently as the sixteenth century.* Some of the very oldest pines at Mar Lodge, the ones greater than 550 years old, may have begun life in a Scotland that was still home to the lynx. But they were absolutely, definitely, 100 per cent extinct by around 1700, since by this point there would not have been enough woodland left to support them.

One of the reasons that they don't feature prominently in our cultural history is that lynx really steer clear of humans. They are distinctly woodland creatures, ambush predators, stalking close and pouncing rather than running their prey down. Roe deer are about the right size for them to ambush safely and successfully in dense woodland. One lynx will take around fifty roe deer a year, given the chance. But lynx are harmless to humans, and are scared of us.

David lists the three things that would be needed for a successful reintroduction: 'Woodland, deer, and positive human attitudes.

---

* This is the date that David personally favours, based on a fifteenth-century cultural reference to lynx hunting roe deer in North Wales. David says, 'If they were in Wales at that point, they would almost certainly have been in the Highlands then.'

There's the potential for a viable population of 400 lynx in the Highlands.'

Where should lynx be reintroduced, I ask?

'There are many places where lynx could live in the Highlands, but the answer is, wherever the sociopolitics allow.'

What of the dangers to sheep? Understandably, the most vociferous opposition to lynx reintroduction is likely to come from sheep farmers. Lynx can and do take sheep, but generally far fewer than people think: the number ranges from zero in countries such as Slovakia to 7,000–10,000 per year in Europe's worst-case scenario by far, Norway.

This difference is attributed to farming practices. Usually, sheep in Norway are grazed in the forest, exactly where the lynx (and a wide range of other predators) live, and where sheep cannot behave as a flock as an anti-predator behaviour. David makes another point, suggesting that a direct comparison between sheep predation in Norway and what we might expect in Scotland is potentially misleading: 'In Norway roe deer densities are very low compared to Scotland, meaning that there are fewer of the lynx's preferred prey species for them to predate on.' He's more keen on looking to Switzerland to see what might happen in Scotland: 'In Switzerland, where there is a flourishing lynx population and where, like Scotland, the majority of sheep are grazed in open pastures and where deer densities are high, fewer than thirty or so sheep are killed by lynx annually.'

Losing sheep in any number is, of course, distressing for farmers. But there may also be benefits for farmers from having lynx around. 'In some landscapes the numbers of foxes killed [by lynx] are high enough to suppose lynx may be targeting them. By killing and displacing foxes in Sweden and Finland, lynx have been shown to benefit capercaillie, black grouse and mountain hares, all species of conservation concern in the UK.'

Is it possible then to generate widespread support from farming communities for lynx? 'I certainly think it's possible. It would require patience and skill and respect. Lynx reintroduction should be done democratically. This is an animal that folks here are very unfamiliar with. It's very understandable that people will have concerns.'

Another thing David points out is that even if there was complete agreement that lynx should be reintroduced to Scotland, actually reintroducing species like lynx can be extremely difficult. The process itself would be torturous. Just finding and trapping enough lynx would be difficult. David estimates that you would need a seed population of around thirty animals, reintroducing four to six each year for several years. Ideally these would be young animals, trapped at the end of winter, and probably coming from somewhere like the Baltic states where there is a large, genetically diverse population and roughly similar conditions to Scotland. 'It was way back in the early 1970s that lynx were first reintroduced to Germany. Many of the first projects were not hugely successful, due to poor planning and public engagement, leading to illegal persecution. In fact, across Europe, a third of lynx reintroduction projects have been unsuccessful, a third successful, and another third are uncertain or ongoing.'

Should we reintroduce lynx to Mar Lodge, I ask?

David laughs. 'The habitat is increasingly there, but Mar Lodge by itself is simply not big enough. You could support maybe one male lynx and a couple of females. That's nowhere near enough for a sustainable population. Lynx need huge landscapes. Unless you have lots of your neighbours onside then it's a non-starter.'

The question here is not one of ecology. Instead, it poses profound questions of society: what does it mean to live alongside large animals? It would take the support of whole communities, and large areas of safety for the species outside of nature reserves, for the lynx to return to these shores. For now, the work of conservationists like David is research and awareness-raising: putting forward not only the facts, warts and all, but also the potential ethical and philosophical rationale for a reintroduction. True rewilding regenerates not only the woodlands, but also our relationships with the woodlands. If we feel that other countries should conserve their tigers and lions and leopards, and we can't even tolerate the idea of living with a shy, Labrador-sized predator of roe deer, then what sort of civilisation are we?

Mar Lodge is more than just a nature reserve. It is a place where we have proved that our relationship with the environment can be sustainable and beneficial for everyone. It is a place of mystery and wonder and excitement. It is a diverse assemblage of ways of life, of interactions between human and non-human cultures, of destruction and regeneration. For now it remains depleted of some of its greatest parts. But it needn't remain so depleted forever.

I have a rule in life: any day that you see an eagle is a good day. March 2020, the world turning upside down, nearly a year on from my minor pilgrimage to the three trees, and I decide to take a walk out to see some eagles. I'm not going to tell you which glen I'm walking up. Eagles are rare, and the locations of their eyries are coveted by egg collectors and rogue falconers.

It's another early start – we're through the dark days of January, when the sun never really seems to rise at all in the glen, and I'm keen to make the most of the returning daytime. There's less wildlife about in March than there is in May. I see little as I walk through the estate. The woods are quietish, with a smattering of coal tits singing. A chaffinch has a go at a little burst of song, a peremptory, throat-clearing, remembering-how-to-sing sort of song. It has been a long winter, and he is clearly out of practice. The snow lies at around 500 metres altitude, following the storms of February. It's a long walk in, through rough ground, but I know the Mar Lodge eagles well and can guess roughly where I might spot one. I walk through the budless, purple-tinged birchwoods, the trees looking poised and ready to explode into leaf at the slightest hint of warmth. Through the pines, where I startle a single mountain hare, still mostly white, dirty flecks of blue-brown beginning to come through on its back. I check on a twinflower patch; its mat of leaves is easier to see at this time of year. I'm not really expecting to see anything particularly interesting, just the dense mat of intertwining leaves and red stems. But it's nice to see it, nevertheless, like visiting a friend you've neglected. Stalks of blaeberry look forlorn. Clumps of cowberry retain their leaves all year round – it is cold, still, and I question their life choices. I hit the woodland edge, and trudge through long heather, onto the moor, still claggy with week-old snow. I find myself a rock with a grand panoramic view and shelter from the wind. I sit down and make myself comfortable. This isn't a raptor watch, and I'm not too worried whether the birds turn up or not, so I let my mind drift.

For over a hundred years, people have come to Mar Lodge and walked out to study the eagles. Seton Gordon himself took the first photograph of a Deeside eyrie, in 1904. Some of the eyries have been in near-constant use for all of that time. Eagles will use and reuse nest sites over successive generations. Every year the eagles are compelled to add a new layer of sticks and twigs to the nests. They grow and grow, reaching ridiculous heights, sometimes metres tall. Sometimes they grow so much that the eagles start bashing their heads against higher branches, at which point they start again on a new eyrie.

In more recent years, satellite tagging data has provided us with previously unimaginable insights into what our eagles get up to and where they go on their long, juvenile wanderings. They have shown us, too, the dark side of our fascination with the species, the shootings and poisonings that stain our relationship with the creatures.

This year, annual monitoring efforts are in doubt. I read the news before heading out, check the ever-increasing restrictions, and the death toll. I speculate as to whether I'll be allowed to walk these glens in a month's time.

For a hundred years, naturalists have come to Mar Lodge to watch the eagles, and added to our collective understanding of them. For hundreds of years before that, people have looked up at the Mar Lodge eagles and our culture has grown with them, stick after stick, an increasingly complex nest of ideas and interactions. A flowery band, binding us to the earth, a fretted dome. Nature is indeed the most complicated word in the language.

What will Mar Lodge look like in a hundred years' time? Two hundred? Five hundred?

Will the pinewoods consolidate their expansion, pushing up higher and further than we thought possible, joined by broadleaves, willows, rowans, aspens and birches? Will willows return to Carn nan Seileach and Allt Coire an t-Seilich, rowans to Allt a' Chaorainn, pines to Glen Geusachan?

I trace a line of regeneration up the hill. The fuzz grows out to the north-east of a patch of granny pines. Research by a student in 2019 found that this is the best aspect for regenerating birch and pine.[10] Will crested tits return to the Deeside pinewoods, I wonder, after a centuries-long absence, scouting through the pines above carpets of twinflower? Will wrynecks, now extinct as a breeding species, return to the pinewoods? What southern species will join them – climate refugees, nuthatch, garden warbler? Will the pines move with deer, stalked not just by humans, but by other, unseen predators?

Will the birches become home to not just redstart and willow warbler and tree pipit, but also redwing, pied flycatcher, brambling? Will there be meadows of alpine sow-thistle, grazed on by deer, rooted out by wild boar with robins and pine seedlings following in their wake? I think back to the Tamworth pigs we kept in an experimental plot in the pinewood, whose rooting around caused an explosion of tree seedling numbers. Will Highland cattle return to the woods, grazing as they did for the crofters of the eighteenth century, filling the role of the extinct aurochs, opening up glades for lekking capercaillie, creating new niches for wildflowers?

For now, the only sound is a mistle thrush, the stormcock, singing its lonely, melancholy song from the nearby pines. After the long months of silence and snowstorms, it is a tonic. I listen greedily to its short, syrupy phrases, never more than a few notes at a time, but endlessly complex. After the long winter, life is returning to the glen.

Will narrow-headed ants once more be part of a functioning ecosystem, gardening the woodland edges? Will pine hoverfly return to the rotting stumps left by foresters, working in harmony with the expanding woods? Will Kentish glories once again follow the newly emerging birchwoods on their wandering journeys across the glens? Will pearl-bordered fritillaries, Scotch argus, northern brown argus continue in their annual emergences, as they have done for centuries? In 2019 they were joined by comma butterflies, the first time they've ever been recorded at Mar Lodge. What else might join them?

A justabuzzard flies out from the left. The peripheral vision in humans is very good at picking out movement – one of those weird evolutionary quirks you hear about in the news from time to time. The bird is a fair distance away, but it's always worth watching raptors. Other birds, infuriated by their presence, are drawn in to attack them. And there's always the chance that you'll be wrong, and it will turn out to be something other than a justabuzzard. The week before, Shaila saw the first of the year's returning hen harriers, a ringtail. One has made it back, at least.

Will hunting hen harriers and peregrines be joined by breeding white-tailed eagles, honey buzzards, red kites? Will goshawks be as easy to find in the Mar Lodge woods as they currently are in the suburbs of Berlin? Will climate change bring hobbies? Montagu's harriers? Marsh harriers?

Will our bogs and wet woodland support not just dunlin and golden plover, but also wood sandpiper, greenshank, green sandpiper? Can we protect what we have from the worst ravages of climate change – our red grouse, snipe, mountain hares? Will curlews continue to sing in the lonely glens, bringing comfort and company to those who hear them?

What will happen down on the Quoich Wetlands? Will beavers be repairing their dams, creating habitats for dragonflies, damselflies, kingfishers, salmon parr, lamprey? Will the lapwings, curlews and redshanks be joined by bugling cranes? Will sixty-pound salmon once again throng the Dee? Will vast beds of freshwater pearl mussels once again clean our water for us? Will they be joined by other creatures we thought lost forever?

I'm drifting off when a monstrous bird stoops around the

shoulder of the mountain. It's a golden eagle – you can tell by the wing shape that it's not a white-tailed eagle – but to be that size it must be the big hen bird that haunts this patch. The cultural cachet of goldies doesn't make sense unless you see them up close. From a distance, they might as well be just another 'justabuzzard'. But they truly are apex predators, with 2-metre-plus wingspans. I'm high up, so I can see down onto her golden head, the gingery mantle that gives the species its name. She's an old bird, so is very dark. She is effortless in the air, barely bothering to flap as she descends from hunting on the plateau. She is the soul of the mountain.

I look up, and here's the male, hundreds of metres up in the air, just a dot. He stoops down, quickly gaining speed. He's still a dot against the steel-grey sky when the hen bird also notices him. A few powerful strokes of her vast wings and she's thinking of joining him. His stoop is vertiginous, almost straight down, and then, a few metres from her, he about-turns, heading straight upwards again. This is skydancing: a mating display. The male is showing off his aerial prowess. He's smaller than the hen, maybe three-quarters her size. He soars up high again, resetting for another display, and in the confusion I've lost the hen bird. Another gut-churning stoop and rise, and he's off, over the moors, point proven, maybe. The show is over for today; nature has half revealed its soul to me, and I must be content with that. Here in the glen at least, far from the madding crowd's ignoble strife, it is a good day.

Will our mountains be carpeted with once-lost willows? Will eagles once again hunt over glens draped with montane scrub? Will high-altitude juniper, pine and birch bring a resurgence of ring ouzels? What of our other lost songbirds: bluethroats, Lapland bunting? Will black redstarts colonise the corries, as they have in Scandinavia? Up on the tops, will we still have snow in August, and dotterel breeding on the heights, and snow buntings, and ptarmigan?

I make my way out of the glen, passing a couple of Munro baggers, ice axes fastened to their bags. Covid-19 has started to bite, but we are still in the early stages of the pandemic and proper social protocols have yet to be codified. We keep our distance, waving awkwardly, acknowledging our good fortune to find ourselves in such a place at such a time. After an odd start, the winter of 2019/2020 has proven to be the snowiest since 2015. The Sphinx is buried deep under tens of metres of accumulated snow. Another pair of walkers, less weighed down with winter mountaineering kit, are doing a low-level walk around the glen. We pass the time of day, again from a solemn distance. They say I must have been up early to be this far up the glen at this hour. I tell them to keep an eye on the skies for eagles. Stay safe, we say.

And what of people? Will a newly reinvigorated network of ecosystems stretching not just across Mar Lodge, but across the whole country, reinvigorate our own relationship with the wild, mitigating the worst effects of climate change, making us richer in spirit and purse? Will proud, tweed-clad sportsmen be hunting red grouse and black grouse through willow scrub, and red deer through the expanding pinewoods? Will they be shooting new quarry species, wild boar, perhaps, and resurgent populations of snipe, woodcock, wigeon and teal? How many people will be enjoying this land, managed in trust for the benefit of all? A hundred thousand a year? Two hundred thousand?

These are the questions that we as a society will have to answer for ourselves. All of these suggestions are well within the realms of possibility, and all have been mooted by conservationists. Some are already becoming reality. Whether or not they happen is up to all of us. Is it the future that we want? Do we want it enough?

March 2020: the world sits in a storm, Events with a capital E again conspire against our species. I return to my home. Outside, a mistle thrush, the stormcock, sings its melancholy song, and it is a tonic.

Afterword: The Thin Green Line

A quick story from lockdown: in July 2020, on furlough, unable to fulfil my usual duties and feeling generally useless and miserable, I cycled out into the hills with the vague intention of looking for rare plants. I spent a few hours squelching through the bogs without seeing anything of much interest, but felt better for the endeavour. On the way back, in a patch of scrubby meadow I've passed hundreds of times before, I spotted a single large, obnoxiously beautiful plant. I skidded off the bike and ran back for a proper look. It was a greater butterfly orchid, a species never before recorded on the reserve. Once common across the UK, the species suffered massive declines in the twentieth century. But there it was. A big, beautiful, rare plant, at once boisterous and delicate, an audacious synecdoche of a landscape in recovery.

The stories in this book are mostly good news, but nothing is immune to the ravages of Covid-19, so we must deal with the bad news. Faced with the horrors of death, illness, lockdowns and job losses, 2020 was the year in which many people rekindled their love of the natural world. But 2020 was also the year in which the thin green line of conservation workers got even thinner, and very nearly breached.

When Covid hit, the UK's environmental charities were ill-equipped to deal with the fallout. There are many complex social and political reasons for this, but this is one of the biggest: we like the *idea* of wildlife, but as a society we can't bring ourselves to pay for it. From 2008 to 2019, UK public spending as a percentage of GDP on biodiversity fell by 42 per cent, from a miniscule 0.038 per cent of GDP to an almost inconsequential 0.022 per cent.[1] In 2018–19, the annual budget of Scottish Natural Heritage, the public body charged with protecting our environment, was less than the wage bill of Celtic Football Club.[2] Since 2010, charity spending on environmental conservation has increased as individuals and organisations like the National Trust for Scotland have scrambled to plug the yawning gap in funding driven by austerity. Despite their herculean efforts, they have come nowhere near to covering the amount lost from the public purse. For this reason, right now, in 2021, the best thing you can do to save the environment is to donate as much money as you can afford to as many environmental charities as you can. In the long run, the best thing you can do is demand much, much more from our elected representatives.

When Covid hit, an overstretched National Trust for Scotland realised that it was facing an existential crisis. A decade of public sector defunding, lost income due to restrictions and lost value on investments put the Trust at serious risk of going under. A fundraising appeal generated £3.4 million, keeping the wolf from the door. The Scottish government stepped up, providing £3.8 million in emergency funding. But it was not enough to stop 232 members of staff from being made redundant in September 2020. Mar Lodge Estate was shielded from the worst by its ability as a large, diverse estate to generate significant income of its own. But even still, two members of staff lost their jobs here.

Covid had other, direct impacts on the environmental restoration work underway at Mar Lodge. Much of the ecological monitoring schedule had to be thrown out, leaving holes in datasets and research that can never be filled. A fair few projects were postponed. But it could have been worse. The regenerating woodlands kept expanding. The ptarmigan still haunted the high hills, and the eagles still hunted them. The curlews still sang on the floodplain, the stags still roared in the southern glens, and the salmon still ran up the Dee. People returned too, once restrictions eased, seeking comfort in nature, in greater numbers than ever before. And a greater butterfly orchid was recorded here for the first time ever.

In the time between writing and publication, there was one other piece of 'big news' that happened that has an indirect but important bearing on some of the stories in this book. In November 2020, the Scottish Government committed to licensing the practice of driven grouse shooting. The proposed licensing is the biggest shakeup of moorland management in decades and will include new restrictions on activities including muirburn and mountain hare culling. In her statement to Parliament, Environment Minister Mairi Gougeon said, 'It is clear to me that we could not continue with the status quo. We all benefit from our natural environment and we all have a responsibility to ensure that it is not only protected but enriched ... Those businesses which comply with the law should have no problems at all with licensing.'[3]

Many were delighted: driven grouse shooting is not a particularly popular activity, with a recent poll finding that over 70 per cent of Scots would like to see it not just licensed but banned outright. Alex Hogg, head of the Scottish Gamekeepers Association, was furious: 'This decision will anger our community. It will not be easily forgotten.' Gamekeepers, he said, are now having to 'cope with never-ending scrutiny and inquiry driven by elite charities with big influence over politicians.'[4] Suffice it to say that a shooting organisation talking of others having 'big influence over politicians' is in the realms of pots, kettles and the various colours thereof. Welcoming the news, Anne McCall, Scotland Director of 'elite charity' the RSPB, said 'the illegal killing of birds of prey; muirburn on peatland soils damaging our vital carbon stores; the mass culling of mountain hares; and the continued use of lead ammunition have absolutely no place in 21st century Scotland ... We believe that what has been announced today is supported by an overwhelming weight of evidence and is entirely proportionate.'[5]

Most people I've told about this far-reaching piece of news have said, simply, 'Oh. I would've thought something like that would have been licensed *years* ago.'

2020 was the year that we all ran out of excuses not to do better by nature. This book is therefore dedicated to the thin green line of people who have given up so much of their lives to protect and restore nature, usually with scant reward, and without whom there would not be a story to write this book about.

With thanks to the folks at Mar Lodge: David Frew, for his support for this project; Shaila Rao, for her tenacity in the face of both years of opposition and years of working with me; Paul Bolton and Kim Nielson, rangers extraordinaires; Chris Murphy, James Allen, Dom Bywater and the stalking team; and the house and estate teams, the unsung heroes. Thanks to the many people who I have leant on heavily in the research and production of this book: Rob Wilson, Paul Ross and the team at St Andrews University; Petra Vergunst, Stewart Taylor, Molly Doubleday, Gabby Flinn, Patrick Cook, Rab Rae, Pamela Esson and the River Dee Trust team, Jos Milner, David Jarrett, Brian Etheridge, Gavin Powell, Aline Finger, Ian Francis, Iain Cameron and David Hetherington. Thanks also to the people who commented on early drafts of this book: Shaila Rao, David Cresswell, Mike Daniels and Matthew Linning. Thanks finally to the team at Birlinn, who made my life very easy in the run up to publication.

With love to Mum, Dad, Amy, Jo, Nana, Grandad and Grandma, Judy, Matthew and Bonnie.

With all my love to Lindsay.

Andrew Painting
January 2021

Notes and References

Website links were live at the time of printing and are given for the convenience of readers, who should be aware that they are owned and controlled by third parties. The publisher has no editorial control over these sites and can accept no liability for their content.

Introduction: Three Trees

1. Watson, A., and Allen, E., 2014 (1984), *The Place Names of Upper Deeside*, Rothersthorpe: Paragon.
2. Watson A., 1983, 'Eighteenth century deer numbers and pine regeneration near Braemar, Scotland', *Biological Conservation*, 25, pp. 289–305. This is a fascinating article, in which Watson uses the Earl of Fife's diaries to estimate the population of deer at Mar Lodge from 1783–92. He finds, 'In the earlier years the Earl often complained about red deer being scarce, and took severe action against poaching. The success of his shooting rose, and deer numbers seen by him increased greatly. He noted young pines in places where none exists today. Most trees alive today date from that time, and deer have prevented regeneration since.'
3. This was part of a nationwide search for veteran and ancient trees. The results of this are fascinating, and can be found on this interactive mapping tool: https://ati.woodlandtrust.org.uk/ (accessed 27 May 2020).
4. Fish, et al., 2007, 'Scottish native pine dendrochronology development: a chronology from Mar Lodge', unpublished AOC Archaeology Report for Historic Scotland and the Forest Research Agency.
5. Queen Victoria was an avid diarist; she managed to fill 141 volumes in her time. In 2012, the royal family made public the full set. They make for fascinating reading, and can be found here: http://www.queenvictorias journals.org/home.do (accessed 26 March 2020).

6 Wightman, A., 2013 (2010), *The Poor Had No Lawyers: Who Owns Scotland (And How They Got It)*, Edinburgh: Birlinn.
7 Watson, F., and Stewart, M., 1996, *Woods and People: A History of the Mar Lodge Estate Woodlands*, Edinburgh: National Trust for Scotland.
8 Laughton-Johnston, J., 2000, *Scotland's Nature in Trust: The National Trust for Scotland and Its Wildland and Crofting Management*, London: Poyser.
9 Quoted in ibid.
10 *Daily Telegraph*, 2002, 'Mystery benefactor revealed', 24 January, https://www.telegraph.co.uk/news/1382530/Mystery-benefactor-revealed.html (accessed 26 March 2020).
11 Quoted in National Trust for Scotland, 2017, Mar Lodge Estate Management Plan 2018–2023, unpublished.
12 'To Nature', Coleridge, S. T., 2010, *The Complete Works of Samuel Taylor Coleridge*, Charleston, SC: Nabu Press.
13 Hayhow, D. B., et al., 2016, *State of Nature 2016*, The State of Nature partnership.
14 Van Dooren, T., 2014, *Flight Ways: Life and Loss in the Age of Extinction*, New York: Columbia University Press.
15 Reinert, H., 2012, 'Face of a Dead Bird: Notes on Grief, Spectrality and Wildlife Photography', *Rhizomes: Cultural Studies in Emerging Knowledge*, Issue 23.

Part One: In the Woods

1 A good introduction to the current state of the Cairngorms' woodlands can be found here: Hetherington, D., 2019, 'Conservation of the mountain woodlands in the Cairngorms National Park', *British Wildlife*, August 2018, pp. 393–400.
2 Steven, H. M., and Carlisle, A., 1959, *The Native Pinewoods of Scotland*, Edinburgh: Oliver and Boyd.
3 https://www.st-andrews.ac.uk/catch/projects/the-scottish-pine-project/ (accessed 2 April 2020).
4 A good starting place, with links to pdfs of some of their work, can be found here: https://www.st-andrews.ac.uk/~rjsw/ScottishPine/publications.html (accessed 26 March 2020).

5   Smout, T. C., 1997, *Scottish Woodland History*, Newbattle: Scottish Cultural Press.
6   House of Lords European Union Committee Sub-committee D, 'Inquiry into the adaptation of agriculture and forestry to climate change: The EU policy response', supplementary memorandum: EU woodland, https://www.parliament.uk/documents/documents/upload/wtd10.pdf (accessed 26 March 2020).
7   Watson, F., and Stewart, M., 1996, *Woods and People: A History of the Mar Lodge Estate Woodlands*, Edinburgh: National Trust for Scotland.
8   Warren, G., et al., 2018, 'Little House in the Mountains? A small Mesolithic structure from the Cairngorm Mountains, Scotland', *Journal of Archaeological Science: Reports*, 18, pp. 936–45.
9   Watson and Stewart, *Woods and People*.
10 For those interested in old maps, the National Library of Scotland has a remarkable series of geo-referenced maps available online here: https://maps.nls.uk (accessed 26 March 2020).
11 Michie, J. G., ed., 1901, *The Records of Invercauld MDXLVII–MDCCCXXVIII*, Aberdeen: New Spalding Club.
12 Cordiner, C. 1780, *Antiquities and Scenery of the North of Scotland*, London.
13 Watson and Stewart, *Woods and People*.
14 Taylor, E., 1869 (2010), *Braemar Highlands: Their Tales, Traditions and History*, Whitefish, MT: Kessinger.
15 Smout, T. C., 1993, *Highlands and the Roots of Green Consciousness*, Edinburgh: Scottish Natural Heritage.
16 http://www.queenvictoriasjournals.org/home.do (accessed 26 March 2020).
17 Steven and Carlisle, *The Native Pinewoods of Scotland*.

1   Scots Pine

1   Burbaitė, L., and Csányi S., 2010, 'Red deer population and harvest changes in Europe', *Acta Zoologica Lituanica*, 20:4, pp. 179–88.
2   Putman, R. J., et al., 1989, 'Vegetational and faunal change in an area of heavily grazed woodland following relief of grazing', *Biological Conservation*, 47, pp. 13–32.

3   Gill, R. M. A., 1992, 'A review of damage by mammals in north temperate forests, 3. Impact on trees and forests', *Forestry*, 65, pp. 363–88.

4   Baines, D., Sage, R. B., and Baines, M. M., 1994, 'The implications of red deer grazing to ground vegetation and invertebrate communities of Scottish native pinewoods', *Journal of Applied Ecology*, 31, pp. 776–83.

5   Petty, S. J., and Avery, M. I., 1990, 'Forest bird communities. Forestry Commission Occasional Paper 26', Forestry Commission.

6   Leopold, A., 1949, *A Sand County Almanac: And Sketches Here and There*, Oxford: Oxford University Press. This is a seminal work in ecological thinking, of similar importance to Carson's *Silent Spring*.

7   Fraser Darling, F., 1969, 'Wilderness and Plenty', The Reith Lectures, http://www.bbc.co.uk/programmes/p00h3xk5 (accessed 1 September 2020).

8   Neil Reid is a veteran of the Cairngorms landscape – a trustee of Bob Scott's Bothy and a key player in maintaining the busy Corrour bothy below the Devil's Point. His Cairngorm Wanderer blog is a fascinating source of Cairngorms history, and much of the information on Malcolm Douglas's experiments was lifted from this blog on the subject:https://cairngormwanderer.wordpress.com/2013/05/26/a-60-year-experiment/ (accessed 26 March 2020).

9   Watson, A., and Francis, I., 2012, *Birds in North East Scotland Then and Now*, Rothersthorpe: Paragon.

10 National Trust for Scotland, 1996, 'Mar Lodge Estate Management Plan', unpublished.

11 Rao, S., 2003, 'Woodland regeneration monitoring report 2003', unpublished.

12 Flagmeier, M., 2008, 'Backmeroff transect monitoring report, 2008', unpublished.

13 Rao, S., 2017, 'Effect of reducing red deer *Cervus elaphus* density on browsing impact and growth of Scots pine *Pinus sylvestris* seedlings in semi-natural woodland in the Cairngorms, UK', *Conservation Evidence*, 14, pp. 22–6.

14 Mar Lodge Independent Review Panel, 2011, 'Report for the Board of the National Trust for Scotland into the management of deer, woodland and moorland at Mar Lodge Estate', http://www.deer-management.

co.uk/wp-content/uploads/2011/11/marlodgefinal_complete.pdf (accessed 11 March 2020).
15 *The Herald*, 2012, 'National Trust comes under fire from conservationists – for not culling deer', 8 April, https://www.heraldscotland.com/news/13053439.national-trust-comes-under-fire-from-environmentalists-for-not-culling-deer/ (accessed 26 March 2020).
16 Rao, S., 2019, 'Woodland regeneration monitoring report 2019', unpublished.
17 Painting, A., and Rao, S., 2020, 'Long-term monitoring of semi-natural woodland at Mar Lodge Estate NNR', *Scottish Forestry*, 74:1, pp. 30–9.
18 Laughton-Johnston, J., 2000, *Scotland's Nature in Trust: The National Trust for Scotland and its Wildland and Crofting Management*, London: Poyser.

2 Green Shield-moss

1 Lorimer, J., 2015, *Wildlife in the Anthropocene: Conservation after Nature*, Minneapolis: University of Minnesota Press.
2 Stewart Taylor, personal communication.
3 Taylor, S., 2010, '*Buxbaumia viridis* in Abernethy Forest and other sites in northern Scotland', *Field Bryology*, 100, pp. 9–14.
4 Taylor, S., 2012, 'Records of *Buxbaumia viridis* growing on new substrates', *Field Bryology*, 107, pp. 21–2.
5 Agnew, J., and Rao, S., 2016, '*Buxbaumia viridis* hotspot survives severe flooding', *Field Bryology*, 115, pp. 19–21.
6 Holá, E., et al., 2013, 'Thirteen years on the hunt for *Buxbaumia viridis* in the Czech Republic: still the tip of the iceberg?' *Acta Soc Bot Pol*, 83 (2), pp. 137–45.

3 Roe Deer

1 Watson, A., and Wilson, J., 2018, 'Seven decades of mountain hare counts show severe declines where high-yield recreational game bird hunting is practised', *Journal of Applied Ecology*, 55:6.
2 Bump, J. K., et al., 2009, 'Ungulate carcasses perforate ecological filters

and create biogeochemical hotspots in forest herbaceous layers allowing trees a competitive advantage', *Ecosystems*, 12, pp. 996–1007.

4 Woodland Grouse

1. Rao, S., and Jones, G., 2019, 'Deeside grouse report, 2019', unpublished.
2. Baines, D., Warren, P., and Richardson, M., 2007, 'Variations in the vital rates of black grouse *Tetrao tetrix* in the United Kingdom', *Wildlife Biology*, 13, Supplement 1, pp. 109–16.
3. White, G., 1789, *The Natural History of Selbourne*, London: Benjamin White.
4. Matthew Holmes has written an excellent introduction to the various reintroductions of caper to Scotland. It can be found here: https://holmesmatthew.wordpress.com/2014/11/21/reintroducing-the-capercaillie-to-scotland-1837-1900/ (accessed 11 March 2020).
5. Weir, T., 1980, *Tom Weir's Scotland*, Edinburgh: Gordon Wright.
6. Ewing, S., et al., 2012, 'The size of the Scottish population of capercaillie *Tetrao urogallus*: Results of the fourth national survey', *Bird Study*, 59, pp. 126–38.
7. 'Mar Lodge Estate Gamebag, 1900–2019', unpublished.
8. Crumley, J., 1991, *A High and Lonely Place: Sanctuary and Plight in the Cairngorms*, London: Jonathan Cape.

5 Kentish Glory

1. Schultz, T. R., 2000, 'In search of ant ancestors', *PNAS*, 97:26.
2. Stevenson, R., and Masson, J., 2015, 'Ants, and their use of *sphagnum* and other mosses', *Field Bryology*, 114, pp. 13–16.
3. Yeomans, E., 2019, 'How to save a rare hoverfly: breed it in hummus pots in the Cairngorms', *The Times*, 29 July.
4. Rao, S., 2004, 'Classic wildlife sites: Mar Lodge Estate, Cairngorms', *British Wildlife*, 16:2, pp. 86–94.
5. Sjöberg, F., 2016, *The Art of Flight*, London: Penguin. See also *The Fly Trap* by the same author.
6. Flinn, G., 2019, 'Species Reports 2019', unpublished.

7   Painting, A., 2019, 'Narrow-headed ant at Mar Lodge: annual monitoring report', unpublished.
8   Stockan, J., 2019, 'Narrow-headed ants at Mar Lodge report', unpublished.
9   https://naturebftb.co.uk/the-projects/narrow-headed-ant/ (accessed 11 March 2020).
10 Bell, J. R., Blumgart, D., and Shortall, C. R., 2020, 'Are insects declining and at what rate? An analysis of standardised, systematic catches of aphid and moth abundances across Great Britain', *Insect Conservation and Diversity*, 2020, Special Issue Article.
11 Conrad, K. F., et al., 2006, 'Rapid declines in common, widespread British moths provide evidence of an insect biodiversity crisis', *Biological Conservation*, 132, pp. 279–91.
12 Fox, R., et al., 2013, *The State of the UK's Larger Moths*, Butterfly Conservation and Rothamsted Research, Wareham.
13 Ibid.

6   On Patrol

1   Cronon, W., 1996, 'The trouble with wilderness: Or, getting back to the wrong nature', *Environmental History*, 1:1, pp. 7–28.
2   Reid, N., 2014, 'The Luibeg woods – aftermath of a blaze', https://cairngormwanderer.wordpress.com/2014/07/14/the-luibeg-woods-aftermath-of-a-blaze (accessed 11 March 2020).
3   Rao, S., Agnew, J., and Bradfer-Lawrence, T., 2017, 'Surface fire fails to promote the natural regeneration of *Pinus sylvestris* on Mar Lodge Estate in the Cairngorms', *Scottish Forestry*, Spring 2017, pp. 26–37.
4   Nicoll, B., 2016, 'Agriculture and Forestry Climate Change report card technical paper: 8, Risks for woodlands, forest management and forestry production in the UK from climate change', https://nerc.ukri.org/research/partnerships/ride/lwec/report-cards/agriculture-source08/ (accessed 11 March 2020).
5   Rao, Agnew and Bradfer-Lawrence, 'Surface fire fails to promote the natural regeneration of *Pinus sylvestris*'.
6   Painting, A., and Rao, S., 2019, 'Regeneration Transects Monitoring Report, 2019', unpublished.

7   Elton, C. S., 1966, 'Dying and dead wood' in *The Patterns of Animal Communities*, John Wiley, New York, pp. 279–305.
8   Lyndhurst, B., 'Rapid evidence review of littering behaviour and anti-litter policies', Zero Waste Scotland.
9   White, M. P., et al., 2019, 'Spending at least 120 minutes a week in nature is associated with good health and well-being', *Scientific Reports*, 9, 7730.
10 Hyat., R. A., 2019, 'The effects of outdoor activity on concentration', thesis in fulfilment of final requirements for the MAED degree, St Catherine University.
11 Carrell, S., 2018, 'Scottish GPs to begin prescribing rambling and bird-watching', *The Guardian*, 5 October.

Part Two: On the Moors

1   Harris, S. J., et al., 2019, 'The Breeding Bird Survey 2018', British Trust for Ornithology, 702, https://www.bto.org/our-science/publications/breeding-bird-survey-report/breeding-bird-survey-2018 (accessed 11 March 2020).
2   Wightman, A., 2013 (2010), *The Poor Had No Lawyers: Who Owns Scotland (And How They Got It)*, Edinburgh: Birlinn.
3   Game and Wildlife Conservation Trust, 2017, *Your Essential Guide to Grouse Shooting and Moorland Management*, Fordingbridge: GWCT.
4   Scottishgamekeepers.co.uk/gamekeeping-facts/economic-benefits.html (accessed 11 March 2020).
5   Fletcher, K., et al., 2010, 'Changes in breeding success and abundance of ground-nesting moorland birds in relation to the experimental deployment of legal predator control', *Journal of Applied Ecology*, 47, pp. 263–72.
6   Whitehead, S., Hesford, N., and Baines, D., 2018, 'Changes in the abundance of some ground-nesting birds on moorland in South West Scotland', research report to Scottish Land & Estates and Scottish Gamekeepers Association, Fordingbridge: GWCT.
7   Game and Wildlife Conservation Trust, 2019, *The Knowledge: Every Gun's Guide to Conservation*, Fordingbridge: GWCT.

8 Somerville-Meikle, J., 2016, 'Grouse shooting: why moorland managers are conservation heroes', *Shooting Gazette*, 20 January.
9 The most sustained polemic on driven grouse shooting is Mark Avery's *Inglorious*: Avery, M., 2015, *Inglorious: Conflict in the Uplands*, London: Bloomsbury. The NGO coalition Revive has its own report on driven grouse shooting here: Tingay, R., and Wightman, A., 2018, 'The Case for Reforming Scotland's Driven Grouse Moors', Revive, https://revive.scot/wp-content/uploads/ReviveReport.pdf (accessed 1 September 2020).
10 MacDonald, B., 2019, *Rebirding*, Exeter: Pelagic.
11 Nicoll, M., 2018, 'Changing tracks: The case for better control of vehicle tracks in Scotland's finest landscapes', Scottish Environment Link, https://www.scotlink.org/files/Changing-Tracks_LINK_Hill tracks_Report.pdf (accessed 10 March 2020).
12 Raptor Persecution UK, 2020, 'Hen harrier shot in North Yorkshire – police appeal for info 5 months later', https://raptorpersecutionscotland.wordpress.com/2020/03/05/hen-harrier-shot-in-north-yorkshire-police-appeal-for-info-5-months-later/ (accessed 10 March 2020).
13 Onekind, League Against Cruel Sports, 2019, 'Untold Suffering: How thousands of animals are trapped, snared and killed to protect grouse shooting for sport', https://www.onekind.scot/wp-content/uploads/Untold-suffering.pdf (accessed 1 September 2020).
14 Raptor Persecution UK, 2019, 'More innocent victims caught in traps set on grouse moors,' https://raptorpersecutionscotland.wordpress.com/2019/06/21/more-innocent-victims-caught-in-traps-set-on-grouse-moors (accessed 10 March 2020).
15 Shorrock, G., 2019, 'The Weight of the Law?' https://community.rspb.org.uk/ourwork/b/investigations/posts/tawny-owl-killed-in-trap-in-trap-north-yorkshire (accessed 10 March 2020).
16 Raptor Persecution UK, 2018, 'Gamekeeper cautioned after merlin killed in illegally-set trap on grouse moor', https://raptorpersecutionscotland.wordpress.com/2018/08/17/gamekeeper-cautioned-after-merlin-killed-in-illegally-set-trap-on-grouse-moor/ (accessed 10 March 2020).

17 Burns, J., 2016, 'RSPB hits out after gull is snared in deadly illegal trap in the Cairngorms National Park', *The National*, 23 July.
18 Raptor Persecution UK, 2017, 'Pine marten caught in spring trap on highland shooting estate', https://raptorpersecutionscotland.wordpress.com/2017/03/16/pine-marten-caught-in-spring-trap-on-highland-shooting-estate (accessed 10 March 2020).
19 Cockburn, H., 2019, '"Sickening": Golden eagle spotted near royal family's Balmoral estate with illegal trap attached to legs', *The Independent*, 14 August.
20 Friends of the Earth, 2019, 'Friends of the Earth sparks moorland burning investigation' https://friendsoftheearth.uk/climate-change/friends-earth-sparks-moorland-burning-investigation (accessed 11 March 2020).
21 Brown, L. E., Holden, J., and Palmer, S. M., 2014, 'Effects of moorland burning on the ecohydrology of river basins. Key findings from the EMBER project', University of Leeds. This enormous piece of work was particularly important, and particularly controversial. A two-page summary of its findings can be found here: https://water.leeds.ac.uk/wp-content/uploads/sites/36/2017/06/EMBER_2-page_exec_summary.pdf (accessed 11 March 2020). From time to time, people interested in continuing the practice of muirburn on peatlands will say that the EMBER project has been debunked. A rebuttal of these claims can be found here: Brown, L. E., et al., 2019, 'Contextualising UK moorland burning studies: geographical versus potential sponsorship-bias effects on research conclusions', https://www.biorxiv.org/content/10.1101/731117v1.full (accessed 11 March 2020).
22 Committee on Climate Change, 2020, *Land Use: Policies for a Net Zero*, https://theccc.org.uk/publication/land-use-policies-for-a-net-zero-uk/ (accessed 1 September 2020).
23 National Trust for Scotland, 2012, 'Wildfires and prescribed burning: policy note', unpublished.
24 Littlewood, N. A., et al., 2019, 'The influence of different aspects of grouse moorland management on non-target bird assemblages', *Ecology and Evolution*, 9:19.
25 Douglas, D. J. T., et al., 2020, 'Benefits and costs of native reforestation

for breeding songbirds in temperate uplands', *Biological Conservation*, 244, 108483.

26 OneKind, 2020, 'Charity urges Scottish Government to introduce legal protection for mountain hares without delay ahead of mountain hare shooting season', https://www.onekind.scot/charity-urges-scottish-government-to-introduce-legal-protection-for-mountain-hares-without-delay-ahead-of-mountain-hare-shooting-season/ (accessed 1 September 2020).

27 Buchan, J., 1925, *John Macnab*, London: Hodder and Stoughton.

28 Young, M., 1989, 'The Rules of the Game: Buchan's John Macnab', *Studies in Scottish Literature*, 24:1.

29 Michie, J., 1908, *Deeside Tales*, Aberdeen: Wyllie.

30 Hunter, J., 2014 (1994), *On the Other Side of Sorrow: Nature and People in the Scottish Highlands*, Edinburgh: Birlinn.

31 Rausing, L., 2019, 'Carbon carnage: the real cost of grouse-shooting', *Standpoint*, 4 December 2019.

32 Grouse Moor Management Review Group, 2019, 'Report to the Scottish Government', https://www.gov.scot/publications/grouse-moor-management-group-report-scottish-government/ (accessed 11 March 2020).

33 Carrell, S., 2019, 'Scottish grouse moors face mandatory licensing', *The Guardian*, 19 December.

7 Red Deer

1 Public and Corporate Economic Consultants (PACEC), 2016, 'The contribution of deer management to the Scottish economy: A report prepared by PACEC on behalf of the Association of Deer Management Groups'.

2 Scottish Natural Heritage, 2016, 'Deer management in Scotland: Report to the Scottish Government from Scottish Natural Heritage 2016'.

3 Wightman, A., 2013 (2010), *The Poor Had No Lawyers: Who Owns Scotland (And How They Got It)*, Edinburgh: Birlinn.

4 MacDonald, B., 2019, *Rebirding*, Exeter: Pelagic.

5 Deer Working Group, 2020, 'The management of wild deer in Scotland', https://www.gov.scot/publications/management-wild-deer-scotland/ (accessed 11 March 2020).

6 Putman, R., Nelli, L., and Matthiopoulos, J., 2019, 'Changes in body-weight and productivity in resource-restricted populations of red deer (*Cervus elaphus*) in response to deliberate reductions in density', *European Journal of Wildlife Research*, 65:13.

7 A good place to start when comparing the health of Scotland's red deer with other countries is the work of Dr Duncan Halley (who also knows a fair bit about beavers). He has written a very accessible presentation on red deer hunting in Norway: nina.no/Portals/NINA/Bilder%20og%20dokumenter/History%20and%20management%20of%20red%20and%20other%20deer%20in%20Norway.pdf (accessed 26 March 2020).

8 Deer Working Group, 'The management of wild deer in Scotland'.

9 Lyons, I., 2020, 'Proposals to reduce Scotland's wild deer population would create culling "free-for-all" gamekeepers warn', *Daily Telegraph*, 29 January.

10 Carrell, S., 2019, 'Eagles need to eat too: Grouse moors take new approach to shooting', *The Guardian*, 11 August, https://www.theguardian.com/uk-news/2019/aug/11/highland-estates-swap-driven-for-walked-up-grouse-shooting (accessed 4 March 2020).

8 Sphagnum Moss

1 https://www.walkhighlands.co.uk/munros/munros-by-rating (accessed 11 March 2020).

2 Heaney, S., 1966, *Death of a Naturalist*, London: Faber.

3 https://app.bto.org/birdfacts/results/bob5120.htm (accessed 11 March 2020).

4 Scottish Natural Heritage, 2003, *Boglands: Scotland's Living Landscapes*, Edinburgh: SNH.

5 Brown, L. E., Holden, J., and Palmer, S. M., 2014, 'Effects of moorland burning on the ecohydrology of river basins. Key findings from the EMBER project', University of Leeds, https://water.leeds.ac.uk/

wp-content/uploads/sites/36/2017/06/EMBER_2-page_exec_summary.pdf (accessed 11 March 2020).

6   National Trust for Scotland (Spaven, D., and Luxmoore, R.), 2016, 'Conserving Natural Capital: The National Trust for Scotland's peatlands', Edinburgh: NTS.

7   Committee on Climate Change, 2020, *Land Use: Policies for a Net Zero*.

8   National Trust for Scotland (Spaven and Luxmoore), 'Conserving Natural Capital'.

9   Committee on Climate Change, *Land Use: Policies for a Net Zero*.

10   MacDonald, B., 2019, *Rebirding*, Exeter: Pelagic.

11   https://www.nature.scot/peatland-action-case-study-whats-connection-between-peat-and-fish (accessed 11 March 2020).

12   Brown, Holden and Palmer, 'Effects of moorland burning'.

13   IUCN UK Peatland Programme, 2011, Research Note: Burning and Peatbogs, https://www.iucn-uk-peatlandprogramme.org/sites/default/files/2019-07/110613%20Briefing%20Burning%20and%20Peatbogs_IUCN%20June%202011.pdf (accessed 11 March 2020).

14   For another take, this GWCT study found that a very limited amount of muirburn (once every twenty years), provided the best balance between carbon sequestration and biodiversity: Marrs, R. H., et al., 2018, 'Experimental evidence for sustained carbon sequestration in fire-managed, peat moorlands', *Nature Geoscience*, 12, pp. 108–12.

15   Committee on Climate Change, *Land Use: Policies for a Net Zero*.

16   Littlewood, N. A., et al., 2019, 'The influence of different aspects of grouse moorland management on non-target bird assemblages', *Ecology and Evolution*, 9:19.

17   Watson, F., and Stewart, M., 1996, *Woods and People: A History of the Mar Lodge Estate Woodlands*, Edinburgh: National Trust for Scotland.

18   Scottish Gamekeepers Association, 2019, SGA Blog: Planned Muirburn and Wildfires Are Not the Same, https://news.scottishgamekeepers.co.uk/2019/04/sga-blog-planned-muirburn-and-wildfires.html (accessed 11 March 2020).

19   Ramage, T., 2019, 'Highland wild fires: muirburn "vindicated"', *Strathspey Herald*, 29 November.

20 Holden, J., et al., 2014, 'Fire decreases near-surface hydraulic conductivity and macropore flow in blanket peat', *Hydrological Processes*, 28:5.
21 National Trust for Scotland, 'Wildfires and prescribed burning: policy note'.
22 RSPB, 2017, 'Moorland birds survey result: some like it wet', https://community.rspb.org.uk/placestovisit/dovestone/b/dovestone-blog/posts/moorland-bird-survey-results-some-like-it-wet (accessed 11 March 2020).
23 BBC, 2018, 'Saddleworth Moor fire is out after more than three weeks', 18 July, https://www.bbc.co.uk/news/uk-england-manchester-44880331 (accessed 11 March 2020).
24 Thompson, P., 'The Saddleworth fire and the importance of restoring our peatland habitats in tackling climate change', https://community.rspb.org.uk/ourwork/b/martinharper/posts/the-saddleworth-fire-and-the-importance-of-restoring-our-peatland-habitats-in-tackling-climate-change (accessed 11 March 2020).
25 Carrell, S., 2019, 'Eagles need to eat too: Grouse moors take new approach to shooting', *The Guardian*, 11 August, https://www.theguardian.com/uk-news/2019/aug/11/highland-estates-swap-driven-for-walked-up-grouse-shooting (accessed 4 March 2020).
26 Evans, R., 2019, 'Grouse-shooting estates face ban on burning of peat bogs', *The Guardian*, 29 October.
27 Committee on Climate Change, *Land Use: Policies for a Net Zero*.
28 https://www.nature.scot/climate-change/taking-action/peatland-action (accessed 11 March 2020).

9 Atlantic Salmon

1 Broadmeadow, S., and Nisbet, T. R., 2004, 'The effects of riparian forest management on the freshwater environment: a literature review of best management practice', *Hydrology and Earth System Sciences*, 8 (3), pp. 286–305.
2 Ibid.
3 Nisbet, T. R., and Thomas, H., 2006, 'The role of woodland in flood control – a landscape perspective' in B. Davies and S. Thompson,

eds, *Water and the Landscape: The Landscape Ecology of Freshwater Ecosystems*', Proceedings of the 14th Annual IALE (UK) Conference, Oxford: IALE (UK), pp. 118–25.

4   Scottish Natural Heritage, https://www.nature.scot/professional-advice/safeguarding-protected-areas-and-species/licensing/species-licensing-z-guide/freshwater-pearl-mussels-and-licensing (accessed 11 March 2020).

5   Trout and Salmon, 2017, 'Salmon fishing on the River Dee', https://www.troutandsalmon.com/features-blog/salmon-fishing-on-the-river-dee (accessed 11 March 2020).

6   Canal and River Trust, https://canalrivertrust.org.uk/enjoy-the-waterways/fishing/angling-histories/angling-heroes/georgina-ballantine (accessed 11 March 2020).

7   Jerome, J. K., 1889, *Three Men in a Boat (To Say Nothing of the Dog)*, London: J. W. Arrowsmith.

8   River Dee Trust, 2019, 'River Dee review 2018–2019', http://www.riverdee.org.uk/f/articles/Dee-Annual-Review-2018-19-compressed.pdf (accessed 11 March 2020).

9   https://atlanticsalmontrust.org/morayfirthtrackingproject/ (accessed 11 March 2020).

10 Hughes, J., 2019, 'Calls for goosander culls are ducking a bigger problem', https://scottishwildlifetrust.org.uk/2019/03/jonny-hughes-calls-for-goosander-culls-are-ducking-a-bigger-problem/ (accessed 11 March 2019).

11 Marine Scotland, 2018, 'Scotland River Temperature Monitoring Network. Topic Sheet 90', Edinburgh: Marine Scotland Science.

12 Nekouei, O., et al., 2018, 'Association between sea lice (*Lepeophtheirus salmonis*) infestation on Atlantic salmon farms and wild Pacific salmon in Muchalat Inlet, Canada', *Scientific Reports*, 8, 4023.

13 River Dee Trust, 2020, 'A million trees to save our salmon', http://www.riverdee.org.uk/news/2020/a-million-trees-to-save-our-salmon (accessed 11 March 2020).

14 Ibid.

15 Halley, D., 2019, 'Two coexisting species: beavers & Atlantic salmon', https://www.youtube.com/watch?v=jaJ7Ky4qRRw Dr Duncan Halley

is an expat Scot who has been living and working in Norway for the last three decades. He is an expert on both beavers and the similarities and differences between Norwegian and Scottish woodlands. As interest in woodland expansion in Scotland has increased, he has become an influential figure. This video is well worth a watch. Halley's principal argument is that, so benign are beavers in Norway, no one has *ever* even thought that they might possibly have an effect on fish stocks.

16   https://atlanticsalmontrust.org/morayfirthtrackingproject/ (accessed 11 March 2020).

10   Curlew

1   Eaton, M. A., et al., 2015, 'Birds of Conservation Concern 4: the population status of birds in the United Kingdom, Channel Islands and Isle of Man', *British Birds*, 108, pp. 708–46.

2   Brown, D., et al. 2015, 'The Eurasian Curlew – the most pressing bird conservation priority in the UK?' *British Birds*, 108, pp. 660–8.

3   Wilson, J., et al., 2014, 'Modelling edge effects of mature forest plantations on peatland waders informs landscape-scale conservation', *Journal of Applied Ecology*, 51, pp. 204–13.

4   Graham Appleton runs the extraordinarily interesting blog Wadertales (wadertales.wordpress.com), in which he brings together pretty much all of the most up-to-date research on waders in Europe. In this instance, Wadertales alerted me to this article: Lindström, Å., et al., 2019, 'Population trends of waders on their boreal and arctic breeding grounds in northern Europe', *Wader Study*, 26:3.

5   Nethersole-Thompson, D., and Nethersole-Thompson, M., 1986, *Waders: Their Breeding Haunts and Watchers*, London: Bloomsbury.

6   Douglas, D., et al., 2014, 'Upland land use predicts population decline in a globally near-threatened wader', *Journal of Applied Ecology*, 51, pp. 194–203.

7   Jarrett, D., et al., 2018, 'Investigating wader breeding productivity in the East Cairngorms Moorland Partnership area using collaborative methods', British Trust for Ornithology, 715.

8   Jarrett, D., et al., 2019, 'Investigating wader breeding productivity in

the East Cairngorms Moorland Partnership area using collaborative methods,' British Trust for Ornithology, 723.
9   Grant, M. C., 1997, 'Breeding Curlew in the UK: RSPB research and implications for conservation', *RSPB Conservation Review*, 11, pp. 67–73.
10 Welsh Wader Ringing Group: https://wwrg.org.uk/species/curlew (accessed 13 February 2020).
11 Grant M. C., et al., 1999, 'Breeding success and causes of breeding failure of curlew *Numenius arquata* in Northern Ireland', *Journal of Applied Ecology*, 36, pp. 59–74.
12 Fletcher, K. L., et al., 2010, 'Changes in breeding success and abundance of ground-nesting moorland birds in relation to the experimental deployment of legal predator control', *Journal of Applied Ecology*, 47, pp. 263–72.
13 Franks, S. E., et al., 2017, 'Environmental correlates of breeding abundance and population change of Eurasian curlew *Numenius arquata* in Britain', *Bird Study*, 64:3.
14 Ausden, M., et al., 2009, 'Predation of breeding waders on lowland wet grassland: Is it a problem?' *British Wildlife*, October, pp. 29–38.
15 Littlewood, N. A., et al., 'The influence of different aspects of grouse moorland management on non-target bird assemblages', *Ecology and Evolution*, 9:19.
16 Laidlaw, R. A., et al., 2015, 'Reducing the impacts of predation on breeding waders using landscape-scale habitat management', DEFRA research report on Project LM0301.
17 Laidlaw, R. A., et al., 2015, 'The influence of landscape features on nest predation rates of grassland-breeding waders', *Ibis*, 157:4.
18 Seymour, A. S., et al., 2003, 'Factors influencing the nesting success of lapwings *Vanellus vanellus* and behaviour of red fox *Vulpes vulpes* in lapwing nest sites', *Bird Study*, 50, pp. 39–46.

11   Hen Harrier

1   Raptor Persecution UK, 2019, 'Certificate of appreciation for northern England raptor forum', https://raptorpersecutionscotland.

wordpress.com/2019/12/22/certificate-of-appreciation-for-northern-england-raptor-forum-nerf/ (accessed 22 May 2020).
2. RSPB, 2014, 'Hen Harrier LIFE+ Project Summary', ww2.rspb.org.uk/our-work/conservation/henharrierlife/downloads/Hen-Harrier-LIFE-summary.pdf (accessed 26 March 2020).
3. BBC, 2016, 'Rare hen harrier chicks hatch at Mar Lodge Estate', https://www.bbc.co.uk/news/uk-scotland-north-east-orkney-shetland-36820021 (accessed 11 March 2020).
4. National Wildlife Crime Unit, 2019, 'Scottish gamekeeper who killed protected birds of prey avoids jail', https://www.nwcu.police.uk/news/wildlife-crime-press-coverage/scottish-gamekeeper-who-killed-protected-birds-of-prey-avoids-jail/ (accessed 11 March 2020).
5. North Yorkshire Police, 2018, 'Gamekeeper convicted of shooting two protected owls', https://northyorkshire.police.uk/news/shot-owls/ (accessed 11 March 2020).
6. NatureScot Commissioned Report 982, 'Analyses of the fates of satellite tracked golden eagles in Scotland', https://www.nature.scot/naturescot-commissioned-report-982-analyses-fates-satellite-tracked-golden-eagles-scotland (accessed 1 September 2020).
7. Murgatroyd, M., et al., 2019, 'Patterns of satellite tagged hen harrier disappearances suggest widespread illegal killing on British grouse moors', *Nature Communications*, 10, 1094.
8. https://www.arcgis.com/apps/opsdashboard/index.html#/0f04dd3b78e544d9a6175b7435ba0f8c (accessed 11 March 2020).
9. Scottish Government, 'Wildlife Crime in Scotland: 2018 Annual Report', https://www.gov.scot/publications/wildlife-crime-scotland-2018-annual-report/ (accessed 13 February 2020).
10. Countryside Alliance, 2020, '"Zero tolerance" for raptor persecution: a joint statement', https://www.countryside-alliance.org/news/2020/1/zero-tolerance-for-raptor-persecution-a-joint-s (accessed 11 March 2020).
11. 'Mar Lodge Estate Gamebag, 1900–2019', unpublished.
12. Ritchie, J., 1920, *The Influence of Man on Animal Life in Scotland*, Cambridge: Cambridge University Press.
13. Watson, A., and Nethersole-Thompson, D., 1981, *The Cairngorms*, Perth: The Melven Press.

Notes and References

14 Watson, A., 2011, *It's a Fine Day for the Hill: Hills, Folk and Wildlife, 1935-62*, Rothersthorpe: Paragon.
15 Dickie, M., 2019, 'Grouse shooting estates hoping for more glorious 2019 season', *Financial Times*, 12 August.
16 Countryside Alliance, '"Zero tolerance"'.
17 Youtube.com/watch?v=DU-Qe1foMy0&t=742s (accessed 11 March 2020).
18 *The Guardian*, 2020, 'RSPB flooded with reports of birds of prey being killed', 15 May, https://www.theguardian.com/environment/2020/may/15/rspb-flooded-with-reports-of-birds-of-prey-being-killed (accessed 22 May 2020).

Part Three: In the Mountains

1 Rockall, A., and Harrower, Y., 2019, *Scottish Mountain Rescue: 2018 Review*, https://www.scottishmountainrescue.org/wp-content/uploads/2019/09/SMR-2018-Review-WEB-020819.pdf (accessed 1 September 2020).
2 https://www.ordnancesurvey.co.uk/blog/2019/01/gbs-longest-linear-walk-without-crossing-a-road/ (accessed 13 February 2020).
3 Macfarlane, R., 2003, *Mountains of the Mind: A History of a Fascination*, London: Granta.
4 Shepherd, N., 2011 (1977), *The Living Mountain*, Edinburgh: Canongate.
5 Macfarlane, *Mountains of the Mind*.
6 Unna, P., 1937, 'The Unna Principles', Edinburgh: National Trust for Scotland.
7 National Trust for Scotland, 2002, 'Wild Land Policy 2002', Edinburgh: NTS.
8 Muir, J., 1901, *Our National Parks*, Boston: Mifflin and Company.

12 Alpine Sow-thistle

1 Hetherington, D., 2019, 'A short history of Scotland's Lost Species 5: The Alpine Blue-sowthistle', https://www.linkedin.com/pulse/

short-history-scotlands-lost-species-5-alpine-blue-david-hetherington?articleId=6586985115476340738#comments-658698511547 6340738&trk=public_profile_article_view (accessed 13 February 2020).
2. Frachon, N., 2017, 'An expedition to the Cairngorms to rescue the rare and endangered Alpine sow-thistle', https://stories.rbge.org.uk/archives/27788 (accessed 1 April 2020).

13   Sphinx

1. https://www.nts.org.uk/stories/new-discovery-on-ben-lawers.
2. Watson, A., and Cameron, I., 2010, *Cool Britannia: Snowier Times in 1580–1930 Than Since*, Rothersthorpe: Paragon.
3. Rivington, M., et al., 2019, 'Snow Cover and Climate Change in the Cairngorms National Park: Summary Assessment', Aberdeen: James Hutton Institute, https://cairngorms.co.uk/resource/docs/boardpapers/06122019/191206CNPABdPaper2Annex1CXC-CairngormsSnowCoverReport-%20FINAL%20-%2022Nov19.pdf (accessed 1 September 2020).
4. http://www.queenvictoriasjournals.org/home.do.
5. Boyes, B., 2017, 'Scottish snow patches: investigating the influences on snow patch distribution and survival', dissertation for the University of Edinburgh.
6. Whitehouse, A., 2015, 'Listening to Birds in the Anthropocene: The Anxious Semiotics of Sound in a Human-Dominated World', *Environmental Humanities*, 6, pp. 53–71.
7. https://www.facebook.com/groups/snowpatchesscotland (accessed 1 September 2020).
8. A line shamelessly stolen from Ted Hughes: Hughes, T., 1974, *Seasons Songs: Spring Summer Autumn Winter*, London: The Rainbow Press.

14   Footpath

1. Naismith, W. W., 1892, 'Excursions. Cruach Ardran, Stobinian, and Ben More', *Scottish Mountaineering Club Journal*, 2:3, p. 136.

2. Newman, T., 'Why we're adjusting Naismith's Rule', 2019, https://www.ordnancesurvey.co.uk/blog/2019/05/why-were-adjusting-naismiths-rule/ (accessed 13 February 2020).
3. Ratcliffe, D., 1977, *Highland Flora*, Inverness: Highlands and Islands Development Board.
4. Painting, A., 2017, 'Removal of the Beinn a' Bhùird track', *Wild Land News*, 91, Autumn, pp. 32–4.
5. Nash, R. F., 2001 (1967), *Wilderness and the American Mind*, New Haven and London: Yale University Press, p. 1.
6. Scottish Natural Heritage, 2019, 'Scottish Upland Path Audit: An initial investigation into the extent of path repairs needed in Scotland's hills', https://www.nature.scot/Scottish-upland-path-audit-report-and-appendices (accessed 10 March 2020).

15 Dotterel

1. Nethersole-Thompson, D., 1973, *The Dotterel*, London: Collins. This book really is the book that started it for all dotterel studies in Britain, so it should come as no surprise that all quotations in this chapter are from this book unless otherwise noted.
2. Thompson, D. B. A., 1991, 'Monitoring the dotterel population of Great Britain', *Britain's Birds in 1989/90: The Conservation and Monitoring Review*, Peterborough: BTO and NCC.
3. Whitfield, D. P., 2002, 'The status of breeding dotterel *Charadrius morinellus* in Britain in 1999, *Bird Study*, 49, pp. 237–49.
4. Hayhow, D. B., et al., 2015, 'Changes in the abundance and distribution of a montane specialist bird, the Dotterel Charadrius morinellus, in the UK over 25 years', *Bird Study*, 62:4, pp. 443–56.
5. Bradfer-Lawrence, T., and Rao, S., 2013, 'Counts of Dotterel and Ptarmigan on the Beinn a' Bhùird plateau, Cairngorms between 2003 and 2012', *Scottish Birds*, 33:3.
6. Ewing, S. R., et al., 2020, 'Clinging on to alpine life: Investigating factors driving the uphill range contraction and population decline of a mountain breeding bird', *Global Change Biology*, 26:7, pp. 3771–87.

16   Downy Willow

1. Ashmole, P., 2006, 'The lost mountain woodland of Scotland and its restoration', *Scottish Forestry*, 60:1, pp. 9–22.
2. Gilbert. D., Horsfield, D., and Thompson., D. B. A., eds 1997, 'The ecology and restoration of montane and subalpine scrub habitats in Scotland', *Scottish Natural Heritage Review*, 83, pp. 21–34.
3. Beck, A., 2016–18, 'Montane scrub in the Cairngorms: report for National Trust for Scotland', unpublished.
4. Keats, J., 1818, *Endymion*, London: Taylor and Hessey.
5. Lintern, D., 2020, 'Highland Haven', *Scotland in Trust*, Spring, pp. 22–8.
6. Shepherd, N., 2011 (1977), *The Living Mountain*, Edinburgh: Canongate.

Epilogue: You Could Have It So Much Better

1. Monroe, M. J., et al., 2019, 'The dynamics underlying avian extinction trajectories forecast a wave of extinctions', *Biology Letters*, 15:12.
2. Coigach & Assynt Living Landscape woodland expansion project: https://coigach-assynt.org/project/woodland-expansion/ (accessed 10 March 2020).
3. Ashmole, M., and Ashmole, P., 2020, *A Journey in Landscape Restoration: Carrifran Wildwood and Beyond*, Dunbeath: Whittles.
4. Thompson, R., 2019, 'Pinewood Managers visit to Glen Ferrick pinewoods – 2 May 2019', *Scottish Forestry*, 72:2, Autumn.
5. Rowe, M., 2018, 'Replanting the Hebrides', *Geographical*, https://geographical.co.uk/uk/uk/item/2816-replanting-the-hebrides (accessed 10 March 2020).
6. cairngormsconnect.org.uk (accessed 10 March 2020).
7. Barkham, P., 2019, 'Rewilding: How Trees for Life are renewing the Highlands', *The Guardian*, 14 December.
8. https://www.Rewildingbritain.org.uk/rewilding/ (accessed 10 March 2020).
9. Hetherington, D., and Geslin, L., 2018, *The Lynx and Us*, Ballintean: Scotland: The Big Picture.

10 Everard, A., 2019, 'Native woodland regeneration in the Cairngorms: a comparison of habitat suitability for Scots pine and birch establishment', thesis submitted to the University of Aberdeen.

Afterword

1 Department for Environment, Food and Rural Affairs, UK (2019), 'UK Biodiversity Indicators 2019', London: DEFRA.
2 NatureScot (2019), 'Annual reports and accounts, 2018–19', Inverness: NatureScot.
3 Ministerial statement: Independent Review of Grouse Moor Management, 26/11/20, Ministerial Statement: Independent Review of Grouse Moor Management (news.gov.scot) (accessed 4/12/20).
4 Scottish Gamekeepers Association, 2020, Chairman's Comment: Grouse Moor Licensing. Scottish Gamekeepers Association News: CHAIRMAN COMMENT: GROUSE MOOR LICENSING (accessed 04/12/20).
5 RSPB, 2020. 'RSPB Scotland responds to Scottish Government statement on Independent Review of Grouse Moor Management', RSPB Scotland responds to Scottish Government statement on independent review of grouse moor management (accessed 04/12/20).

Index

Abernethy 7, 10, 18, 57, 247, 250
acronyms, the ubiquity of 2, 79
Albert, Prince Consort 1, 9
alder *Alnus glutinosa* 36, 145, 184
Alfred, Lord Tennyson 1
alpine sow-thistle *Cicerbita alpina* 197–206
ammunition, lead 69
An Sgarsoch 17, 107, 130–2
arctic-alpine plants 15, 207–10, 234
Atholl Estate 5, 8, 126, 218
Allt a' Chaorainn 35, 262
Anthropocene 19, 129, 215, 235
aspen *Populus tremula* 25, 28, 38, 145, 184, 204, 262
Atlantic salmon *Salmo salar* *see* salmon

Balfour, Professor John 218
Ballochbuie 34, 250
Balmoral 9
Balmorality 34, 122
beaver *Castor fiber* 10, 155 168n, 251, 253, 255, 263
Ben Avon 17, 231
Ben Lawers 11, 208–9, 239, 247

Beinn a' Bhùird 17, 108, 191, 223, 228, 231
Beinn a' Chaorainn 35
Beinn a' Ghlo 17
Beinn Bhreac 191, 223
Beinn Bhrotain 17, 108, 132, 190–1, 223
Beinn Eighe 10, 250
Beinn Mheadhoin 7, 17, 38, 223
Ben Macdui 2, 16–7, 108, 189, 192, 207, 213
Ben Nevis 190
birch *Betula* 5–6, 17, 26, 31, 35, 38, 43–4, 49, 113, 116, 127, 154, 184, 200, 203–5, 228, 237–8, 248, 262–4
    downy *B. pubescens* 145, 208, 237
    dwarf *B. nana* 135, 144–5, 148–9, 155, 241
    hybrid *B. x intermedia* 208
    importance for Kentish glory 87–9
Black Bridge 5, 127
black grouse *Tetrao tetrix* 2, 9n, 42, 73–82, 93, 99, 110, 149, 255, 258

blaeberry *Vaccinium myrtillus* 4, 25, 50, 64, 80, 154, 244, 261
    See also northern blaeberry *Vaccinium uliginosum*
Bod an Deamhain 17, 207
bog, blanket 30, 109, 111, 134, 137, 141, 159, 184
Botanical Society of Britain and Ireland (BSBI) 208–9
botany 53, 199, 201, 207–10, 231, 248
    Victorian 210, 212, 217–8
bothy
    Bob Scott's 100
    Corrour 210, 216
    Hutchison's Memorial 15, 244
Braemar 2, 9n, 18, 85, 120, 167, 171, 189, 250
Braemar Gathering 122
Braeriach 17, 130, 190, 207, 212, 215–6, 219, 223
British Association of Shooting and Conservation (BASC) 183–4
British Trust for Ornithology (BTO) 107, 157, 160, 162, 165n, 176
brown trout *Salmo trutta* see trout
bryology 53–62
Buchan, John 111–17
Butterfly Conservation 88, 93
buzzard *Buteo buteo* 6, 180–2, 263–4
Bynack Burn 108
Bynack Lodge 107

Bynack More 17
Byron, Lord viii

Cairn Gorm 17, 210
Cairn Toul 17, 109, 207
Cairngorms 2–3, 5, 14–15, 17, 44, 108, 124, 134, 158–9, 189, 196, 203, 228–30, 234, 236, 253
    danger of 192
    montane scrub in 239–48
    moors of 127
    pinewoods of 25–8, 43
    raptor persecution in 179, 182
    rounder delights of 130
    size of 191, 233
    snow patches of 211–6
    Wild Land Area (WLA) 194
Cairngorms Connect 18, 241, 250
Cairngorms National Nature Reserve 41
Cairngorms National Park 2, 80–2, 197, 200
Cairngorms National Park Authority (CNPA) 79, 83, 142, 158, 212, 256
Caledonian Pinewood 4, 5, 7, 16–17, 25–36, 39, 44, 49, 51, 58, 71, 76–7, 99, 109, 113, 185, 243, 253
Caledonian woodland see Caledonian pinewood
Cameron, Iain 212–6
capercaillie *Tetrao urogallus* 25, 42, 55, 75–82, 92, 100, 103, 162, 228, 255, 258, 262

Index

carbon 72, 102, 137–43, 148–9, 251, 254, 257
Carn a' Mhaim 6, 17, 64, 191, 223
Carn an Fhidhleir 17, 130–2, 143
Carn Bhac 17, 235
Carn Crom 100, 102, 191, 223
Carn Ealer *see* Carn an Fhidhleir
Carrifran 250
cattle 9, 31, 95, 131, 153, 262
charisma, non-human 54–5, 61, 76, 81
Clais Fhearnaig 36, 102
Clearances 6, 7, 9, 34, 77, 96, 110, 114, 115, 122, 240
climate change 30–1, 79, 92, 102, 108, 111, 128, 136, 142, 153–4, 183, 206, 212, 224, 233–5, 240, 248, 263, 265
climatology 26–9
Cnapan nan Clach 107–8
Coire an Lochain Uaine 189
Coire an t-Sneachda 210
Coire Etchachan 14–5, 213, 244
Coleridge, Samuel Taylor 15
Committee on Climate Change (CCC) 142
common buzzard *Buteo buteo see* Buzzard
common crane *Grus grus see* crane
common gull *Larus canus* 111, 157
common snipe *Gallinago gallinago see* snipe
conservation 6, 18–22, 64, 71, 77, 79, 82, 86, 92, 136–7, 144, 147–8, 155, 162, 165, 175, 179, 201, 225, 228, 249, 256, 259
   the changing nature of 20, 242, 250–1
   community-led 80–1
   criticism of 44–5, 47–8, 80, 247
   difficulties of 19, 26, 88–9, 91, 153, 154, 195, 246
   funding for 54, 79, 241
   in the Highlands 10–2
   landscape scale 13, 41, 93, 168, 247
   philosophical nature of 18–20, 72, 96, 193–4, 202, 205, 215
   politics of 43–51, 57, 219–20
   and sport 10, 12–13, 22, 47, 51, 64, 110, 128, 153
   see also rewilding
Conservative Party 111, 114, 141
Cordiner, Rev. Charles 32
Corgarff 136, 191
Corrie Fee 247
Corriemulzie Cottage 121
Corrour bothy *see* bothy
Corrour (Rannoch) 199
Countryside Alliance 183
cow *see* cattle
crane *Grus grus* 168n, 251, 263
cranefly *Tipulidae* 141, 229, 233
Creag Bad an-t-Seabhaig 190
Creag Bad an Eàs 9
Creag Fhiachlach 240
Creag Meagaigh 10

crofting 5–6, 9, 31, 34n, 77, 96, 100, 114, 262,
crow *Corvus spp.* 110, 163–7, 255
curlew *Numenius arquata* 107, 110, 141, 157–70, 263

Dalvorar 109, 137, 162
deadwood 6, 43, 50, 55–6, 61, 99
deer
    browsing impacts of 4–5, 34, 40, 44, 80, 123, 144–5, 161, 200, 208, 236–9
    economics of 123–5, 221
    fecundity 125
    fencing 41–2, 48, 79, 204, 247–8
    forest 8n, 109
    hunting *see* deer stalking
    hummel 126–7
    impact of on Highland culture, land management and history 9–11, 39, 51, 65, 69, 110–17, 119–29, 265
    larder 71–2
    management 38, 42–50, 60–72, 117, 119–29, 135, 142, 243, 255
    population size of 4, 10, 40, 42, 63, 124–5
    relationship of with woodland 39, 71, 127
    rut 93, 119–21
    stalking 6–9, 39, 60–72, 110–3, 119–29, 219, 230
    vacuum 45, 47, 70
    watching 121
    see also red deer *Cervus elaphus*, roe deer *Capreolus capreolus*
Deer Commission 44–5
Deeside 1, 4, 30, 33–4, 42, 75, 76, 78, 84, 88, 95, 114, 137, 152, 154–5, 179, 182–3, 240, 250, 261
Derry Cairngorm 6, 16–18, 38, 189–90, 223
Derry Dam 7, 67
Derry Lodge 7, 100
Devil's Point *see* Bod an Deamhain
dogs, impact on rare species 80, 234
Doire Bhraghad 1, 66
dotterel *Charadrius morinellus* 211, 226–35, 264
Dovestone 140–1
Dubh Ghleann 27
Duchess of Argyll, Princess Louise, 9, 111
Duchess of Fife, Alexandra, Princess, Second Duchess 9
Duke of Atholl 218
Duke of Fife, William Duff 9
Dundreggan 245, 248, 250
dunlin *Calidris alpina* 132–43, 163, 263

Earls of Fife 100
    Duff, William, First Earl of Fife, Lord Braco 9
    Duff, James, Fourth Earl of Fife, 77–8

Earl of Mar, John Erskine 8–9
East Cairngorms Moorland Partnership (ECMP) 158, 162, 165n, 166–8
Easter Charitable Trust 12–13, 42
ecology 18–19, 30, 39, 72, 85, 126, 128, 164, 168, 194
ecosystem 13, 18, 40, 71–2, 79, 85, 87, 92, 99, 125, 144, 153, 154, 156, 164, 193, 229, 233, 238–40, 244
    restoration 18, 94, 253–6
    services 93, 149, 251, 265
ecotone 239–40
edge effect 163
Edward VII, King 9
eel *Anguilla anguilla* 152
egg collecting 158, 182, 260
electrofishing 150–1
Elrigs 70
Endangered Landscapes Programme 241
Eurasian curlew *Numenius arquata* see curlew
Eurasian lynx *lynx lynx* see lynx
European eel *Anguilla anguilla* see eel
European larch *Larix decidua* see larch
European Union 43, 57
Ewing, Fergus MSP 44–5
extinction viii, 3, 18, 63, 72, 77–80, 114, 158, 162, 180, 240, 249–50, 253, 256–7, 262

Facebook 216

Farquhar Munro, John MSP 45
field sports *see* deer, gamekeeping, grouse, Highland sport, salmon
Fife Arms 85
fire *see* muirburn, Mar Lodge fire, wildfire
flatfly 94
Flow Country 163
footpaths 14, 80, 160, 217–25
Forest and Land Scotland (FLS) 79, 242
    See also Forestry Commission
forestry 6–9, 31–3, 77, 123, 139, 254
    plantations 6, 30, 78–80, 102, 110, 144, 163, 254
Forestry Commission 70, 78
fox *Vulpes vulpes* 60, 69, 72, 79, 110, 159, 163–7, 255, 258
Fraser Darling, Frank 40
freshwater pearl mussel *Margeritifera margeritifera* 144–9

Gaelic 1n, 15n, 70n, 75, 107, 108, 210, 240, 244
Gallows Tree o' Mar 97
Game and Wildlife Conservation Trust (GWCT) 13, 140
gamekeeping 9–10, 45–6, 48, 51, 110, 114, 115, 117, 120, 124–5, 185, 218, 245
    and woodland restoration 64–70
    and raptor persecution 179–84

and wader conservation 160–6
Garbh Choire emergency shelter 207
Garbh Choire Mòr 209–16
genetics
   of alpine sow-thistle *Cicerbita alpina* 200–3
   of capercaillie *Tetrao urogallus* 82
   of downy willow *Salix lapponum* 245
   of hen harrier *Circus cyaneus* 176
   of narrow-headed ant *Formica exsecta* 91
   of whortle-leaved willow *Salix myrsinites* 245
Ghillies' Ball 122
glacier 7, 145, 207
Glen Affric 250
Glen Dee 10, 108, 114, 144, 162, 191, 223
Glen Derry 7, 32, 38, 41, 66–7, 78, 98–9, 191, 227, 242, 245
Glen Ey 9n, 18, 114, 122
Glen Geldie 17, 30, 44, 108, 109, 132, 135, 137, 144–6, 153–5
Glen Geusachan 31, 109, 262
Glen Lui 5, 7, 32–3, 77, 99, 114, 122, 127
Glen Luibeg 10, 32, 95, 98–9, 190, 191, 223
Glen Quoich 10, 17, 26, 31, 32, 70, 228, 230
Glen Strathfarrar 199, 250
Glen Tilt 2, 108, 217–8
Glencoe 11, 193
Glenfeshie 18, 45–8, 108, 191, 247, 250
globeflower *Trollius europaeus* 49, 199, 208, 244
golden eagle *Aquila chrysaetos* 6, 63, 67, 111, 133, 164, 171, 173, 180, 182, 224, 227, 255, 260–1, 264–5
golden plover *Pluvialis apricaria* 17, 108, 111, 131, 135, 162, 163
goosander *Mergus merganser* 153
Gordon, Seton 40, 212, 261
goshawk *Accipiter gentilis* 133, 173, 180, 181–2, 251, 255
Grampians 18, 124, 230, 234
great sundew *Drosera anglica* 208
green shield-moss *Buxbaumia viridis* 53–62
grey wagtail *Motacilla cinerea* 243–4
grouse
   driven shooting 115, 169, 178, 180–2
   moor management 10, 12, 34, 67, 109–11, 116–17, 139, 165, 167, 177, 179–80, 183–5, 221
   walked-up shooting 42–3, 69, 112, 180–2, 265
   see also black grouse *Tetrao tetrix*, capercaillie *Tetrao urogallus*, muirburn, predator control, ptarmigan

Lagopus mutus, red grouse *Lagopus lagopus*

hare *see* mountain hare *Lepus timidus*
heather *Calluna vulgaris* 4, 12, 17, 43, 50, 80, 83, 110, 112, 115, 127, 133–5, 139–42, 145, 148, 172–4, 252
hen harrier *Circus cyaneus* 6, 114, 171–85, 263
Heritage Lottery Fund (HLF, NHLF) 12, 79, 91
Highland Clearances *see* Clearances
Highland
    sport 4, 9, 11, 51, 71, 109–18, 123, 124–9, 184–5, 250
    sporting estate 4, 8–13, 51, 114–6, 124, 141, 158, 181, 250
Highland Wildlife Park 86, 253
Holocene 29
honey buzzard *Pernis apivorus* 182, 263

Invercauld Estate 46, 250
Inverey 120

James Hutton Institute 91
John Muir Trust (JMT) 146, 245
Julius Caesar 144
Juniper *Juniperus communis* 15, 25, 49, 238, 241–5

Keats, John 246

Kentish glory *Endromis versicolora* 83–94, 166, 263
Kluge, John 9, 11

Lairig Ghru 27, 213, 216, 219
    race 27
Land Reform Act 2003 99
lapwing *Vanellus vanellus* 53, 157, 159–63, 166, 263
larch *Larix decidua* 34, 107, 154
Leopold, Aldo 40
Linn of Dee 17, 32, 97, 120, 149, 154–5, 223
Lister, Paul 129
Loch Avon 16
Loch Etchachan 14, 15, 213, 243
Lochnagar 16, 18
Luibeg Cottage 100
lynx *Lynx lynx* 10, 49, 120n, 164, 252–3, 255–9

Macfarlane, Robert 192–3
Mar Estate 9
Mar Forest 2, 66
Mar Lodge
    ballroom 121–3, 126–9
    building 1, 111
    fire 121n
    flooding 121n
    independent review 47–8, 52
    National Nature Reserve vii, 2, 49
    policy woodland 53
    stables 1
    tweeds 69, 125

Mary, Queen of Scots 8, 119
meadow pipit *Anthus pratensis* 131, 145, 232
merlin *Falco columbarius* 111, 131, 175
midge 37, 52, 94, 99
mistle thrush *Turdus viscivorus* 262, 265
montane scrub 195, 239–48, 250, 264
Moorland Association 183
moorland zone 42, 117, 124, 144
Moray Firth 16, 156
Morrone
    birkwood 250
    footrace 18
moths 83–94
    captive rearing 90
    decline 92–3
    peculiarity of nomenclature 86–7
    trapping 91–2
mountain bikes 221, 233
mountain hare *Lepus timidus* 15, 16, 66–7, 110–11, 145, 149, 204, 211, 243, 258
Mountain Rescue Service 16, 222–3
Mounth 18, 226
Muir, John 196
Muirburn 110–1, 139–42, 167–9, 254
Munros 2, 7, 17, 109, 130, 219, 236

Naismith's Rule 217–18

narrow-headed ant *Formica exsecta* 83–4
National Biodiversity Network (NBN) 59
National Heritage Memorial Fund 12
National Trust for Scotland viii, 20
    agreement with Easter Trust 12
    carbon stores 138
    landholdings 11, 123, 193, 225
    management objectives 12–14, 38, 42, 45–9, 69, 254
    muirburn policy 141
    open access policy 14
    purchase of Mar Lodge 11
    relationship with landed gentry 11n
    rewilding policy 254
    surprising radicalism of 42, 141, 218
    vehicle track restoration 219
    wild land policy 96, 193–5
Nature Conservation Committee (NCC) 41n
NatureScot 10, 41, 57, 79, 142, 205, 221, 227 *see also* Scottish Natural Heritage
nest finding 132–6, 157–60, 169, 172–5
Nethersole-Thompson, Desmond 163, 182, 226–30, 233, 235
northern blaeberry *Vaccinium uliginosum* 15, 239, 244

northern goshawk *Accipiter gentilis see* goshawk
northern lapwing *Vanellus vanellus see* lapwing

Ordnance Survey 191, 217–18
osprey *Pandion haliaetus* 136, 148, 180
oystercatcher *Haematopus ostralagus* 157, 160, 162

Panchaud brothers 9
pearl-bordered fritillary *Boloria euphrosyne* 4, 86, 263
Pearls in Peril Project 148–9, 154
peat 30–1, 98–9, 108, 131, 137–43
Peatland Action Programme 142
People's Postcode Lottery 241
peregrine *Falco peregrinus* 14, 113, 117, 133, 175, 180–1, 190
phenology 93–4
pine *see* Scots pine *Pinus sylvestris*
pine hoverfly *Blera fallax* 85–6
pine marten *Martes martes* 72, 76, 79, 111, 114, 161, 163, 251, 255
Pinnacles (snow patch) 212, 215–6
pipit *Anthus spp. see* meadow pipit *Anthus pratentsis*, tree pipit *Anthus trivialis*
Povlsen, Anders 45, 129
Preas nam Meirleach 95
precautionary principle 246–7
predators
    apex 72, 164, 252–60, 264
    control of 110–11, 117, 153, 164–7, 169
    ecological impacts of 39–40, 63, 79, 87, 148, 153, 159, 161, 163–4, 173, 182, 256
    generalist 79, 159, 163–4, 255
    reintroduction of 252–60
Princess Alexandra *see* Duchess of Fife
Princess Louise *see* Duchess of Argyll
ptarmigan *Lagopus mutus* 9, 16, 190, 211, 231–2, 264

Quoich Flats 17, 120, 157, 162, 167–9, 263
Quoich Water 17, 157, 160

Rao, Shaila 37–52, 53, 60, 61, 64, 65, 75, 189, 230, 236–7, 245–6, 263
Ramblers 156, 218
rangers 70, 80, 95–103, 158, 160, 225
raptor persecution 110, 171, 178–83
rare invertebrates in the Cairngorms 83, 85, 91
Rausing, Lisbet and Sigrid 129
red deer *Cervus elaphus* 4–5, 10, 34, 39–40, 70–1 119–29, 161, 265
see also deer
red fox *Vulpes vulpes see* fox
red grouse *Lagopus lagopus* 2, 34, 110–11, 115, 145, 165, 181, 184–5, 190, 263, 265 *see also* grouse

red kite *Milvus milvus* 29, 180, 182, 251, 255, 263
red squirrel *Sciurus vulgaris* 1, 76, 251, 255
redshank *Tringa totanus* 157, 162, 263
regeneration zone 42, 45, 48, 66, 144, 169, 242
reindeer *Rangifer tarandus* 15, 145, 213
River Dee 2, 12, 17, 149, 150, 152–6
River Dee Trust 150–5
River Tay 149, 155, 156
rewilding 20, 72, 251–5, 259
Rewilding Britain 253–4
ring ouzel *Turdus torquatus* 111, 145, 209, 231, 264
Robber's Copse *see* Preas nam Meirlecah
roe deer *Capreolus capreolus see also* deer 49, 63–72, 208, 252–3, 255–8
Romans 25–6, 144
Romanticism 33, 192
Rothamsted Research 92
Rothiemurchus 96, 255
rowan *Sorbus aucuparia* 15–16, 31, 35, 49, 145, 148, 199, 237–8, 240, 243–4, 262
Royal Botanic Garden Edinburgh (RBGE) 197–201, 245
Royal Society for the Protection of Birds (RSPB) 13, 57, 79, 83, 140–1, 162, 164, 176, 178, 180, 182, 183

Saddleworth Moor 140–1
Salmon *Salmo salar* 42, 69, 110–13, 144–56, 263
Salvesen, Ann Marie 12
Salvesen, Andrew 13
Scandinavia 65, 67, 78, 82, 126, 133, 145, 163, 167, 200, 202–3, 206, 221, 226, 233, 258, 264
Scheduled Ancient Monument (SAM) 3, 114
Schedule One species (birds) 173, 228
Scots pine *Pinus sylvestris* 3–8, 13–14, 37–52, 66, 80, 84–5, 113, 144, 163–4, 203, 204, 228, 238, 262, 264
   carbon store 137
   coring 27–36
   granny 3, 41, 100, 262
   importance as a keystone species 39
   Krummholz 238–9, 243–4
   longevity in landscape 25–32, 108, 131, 138, 257
   relationship with deer 34, 39–41, 64, 71
   use by people 5–10, 31–4, 77–8, 140
   see also Caledonian pinewood
Scott, Bob 182
Scottish crossbill *Loxia scotica* 7, 227
Scottish Gamekeepers Association (SGA) 46, 51, 128
Scottish government 40, 116, 183

Index

Scottish Outdoor Access Code 14, 99, 220
   See also Land Reform Act 2003
Scottish Forestry (SF) 79
Scottish Natural Heritage (SNH) 10–11, 45–6, 123, 194, 221
   see also NatureScot
Scottish Raptor Study Group 175
Scottish Rights of Way and Access Society 218
Scottish Semi-natural Woodland Inventory 239–40
Second World War 32
Sentinel satellite 213
Sgor an Lochain Uaine 17, 207
Shakespeare, William 95
sheep 157, 161, 164, 239, 258
Shepherd, Nan viii, 192, 213, 248
slug 204–5
small cow-wheat *Melampyrum sylvaticum* 145, 199
snipe *Gallinago gallinago* 131, 157, 162, 263
snow 2, 6, 15, 53, 65–7, 71 189–91, 227, 236, 260, 264–5
   cultural significance of 16, 210–11, 214–6
   ecological impacts of 98, 134, 145, 153, 161, 211, 233, 237–8, 243–4
   patches 207–16, 236
snow bunting *Plectrophenax nivalis* 190, 209, 211, 213, 231, 264
Speyside 78, 80, 83, 106, 136, 156, 240, 250

Sphagnum moss 130–43
Sphinx (snow patch) 207–16, 217, 265
sport *see* conservation and sport, deer, grouse, Highland sport, salmon
Sputain Dearg 17, 189–90
stag *see* deer, red deer *Cervus elaphus*
Stag Ballroom *see* Mar Lodge ballroom
stalking (deer) *see* gamekeeping
Storm Frank 154
Strathspey *see* Speyside

temperature logger 153, 159–61
Thistle Camp 225
tree pipit *Anthus trivialis* 90, 93
Trees for Life 245
trout *Salmo trutta* 147, 150, 151–2
twayblade *Neottia spp.* 49, 50
twinflower *Linnaea borealis* 145, 198–9, 260

University of the Highlands and Islands (UHI) 124
Unna, Percy 193–5
Upland Footpath Advisory Group 221

Victoria, Queen 1, 7, 9, 33, 121n, 122, 213
Victorians 34, 56, 58, 210, 217–18,
vole *Microtus spp.* 204
   appearance in lunch box 132

population cycle 173, 177
  importance as a prey item 165, 172

Watson, Adam 1n, 3–4, 15n, 48, 182, 211–12, 227
Werritty Review 116
western capercaillie *Tetrao urogallus* see capercaillie
white-tailed eagle *Haliaeetus albicilla* 71, 114, 180, 182, 251, 255, 264
whitebeams, the Arran *Sorbus spp.* 199
wild boar *Sus scrofa* 10, 255, 262
wild camping 100, 154, 190
wild land 20, 96, 130, 193–5, 220–1, 224
Wild Land Areas (WLA) 3, 194
wildfire 97–9, 140
willow *Salix spp* 5–6, 28, 49, 66, 145, 148, 168, 184–5, 236–48, 262
  creeping *S. repens* 148, 238
  downy *S. lapponum* 236–48
  eared *S. aurita* 148, 238, 242–5
  grey *S. cinerea* 238, 242
  hybridisation 245
  montane scrub 15, 199, 211, 236–48
  mountain *S. arbuscula* 244–5
  tea-leaved *S. phylicifolia* 208, 242
  whortle-leaved *S. myrsinites* 242, 245
willow warbler *Phylloscopus trochilus* 2, 90, 134, 229
Windsor Great Park 127
wolf *Canis lupus* 7, 8, 10, 39, 49, 120n, 164, 252, 255–7
wood ant *Formica spp.* 43, 50, 60, 83–4
woodland regeneration 10, 13, 34, 37–52, 63–72, 90, 100, 124, 144, 241, 250
Woodland Trust 4